Chemoton Theory

Volume 2: Theory of Living Systems

MATHEMATICAL AND COMPUTATIONAL CHEMISTRY

Series Editor: PAUL G. MEZEY
University of Saskatchewan
Saskatoon, Saskatchewan

CHEMOTON THEORY
Tibor Gánti

FUNDAMENTALS OF MOLECULAR SIMILARITY
Edited by Ramon Carbó-Dorca, Xavier Gironés, and Paul G. Mezey

MANY-ELECTRON DENSITIES AND REDUCED DENSITY MATRICES
Edited by Jerzy Cioslowski

SIMPLE THEOREMS, PROOFS, AND DERIVATIONS IN QUANTUM CHEMISTRY
István Mayer

Chemoton Theory

Volume 2: Theory of Living Systems

Tibor Gánti

Lóránt Eötvös University
Budapest, Hungary

Kluwer Academic / Plenum Publishers
New York • Boston • Dordrecht • London • Moscow

Library of Congress Cataloging-in-Publication Data

Gánti, Tibor–
[Chemoton elmélet. English]
Chemoton theory/Tibor Gánti.
p. cm. — (Mathematical and computational chemistry)
Includes bibliographical references and index.
Contents: v. 1. Theoretical foundations of fluid machineries — v. 2. Theory of living systems.
ISBN 0-306-47785-8 (set) — ISBN 0-306-47783-1 (v. 1) — ISBN 0-306-47784-X (v. 2)
1. Life—Origin. I. Title. II. Series.

QH325.G355 2003
576.8′3—dc21

2003050647

Translated from the original Hungarian versions (1984 and 1989) by Elisabeth Czárán.

Volume 1: ISBN 0-306-47783-1
Volume 2: ISBN 0-306-47784-X
Set: ISBN 0-306-47785-8

233 Spring Street, New York, New York 10013

http://www.wkap.com

10 9 8 7 6 5 4 3 2 1

A C.I.P. record for this book is available from the Library of Congress

Printed in the United States of America

Contents

VI. About the World of Living in General **251**
What is Life? 251
The ambiguity of the concept of life 251
Views concerning the essence of life 254
The standpoints of some outstanding philosophers 259
The Most General Properties of Living Beings 268
Empirical facts providing the basis for the theory 268
Conclusions 273
Life Criteria 278
Alive, capable of living, dead, nonliving 278
The levels of life 280
Life criteria 282
Chemotons as Living Systems 289
Chemotons satisfy life criteria 289
Chemotons are minimum systems of life 293

VII. Chemical Evolution **297**
Setting of Objectives 297
The Date of the Genesis of Life 298
The Genesis of Earth 300
The Primitive Atmosphere 308
Chemical Evolution 311
Introduction 311
Energy source 312
Initial compounds 314
Low-molecular-weight products 315
Abiotic formation of proteinoids 324
Abiogenic formation of nucleic acids 325
Formation of microscopic structures 328

VIII. Biogenesis **331**
Models of Biogenesis 331
Experimental models 331
Qualitative theoretical models 332

Quantitative theoretical models 334
Metabolic Network of a Prebiotic Chemoton 338
General aspects 338
Autocatalytic subsystems 340
Connective subsystems 346
Additional pathways 351
A qualitative survey of the operation of the network 354
Quantitative Description of Prebiotic Chemotons 356
Strategy 356
Membrane synthesis 359
Template synthesis 361
Combination of the two subsystems 371
Synthesis of precursors 373
Self-reproducing metabolic network 377
Overall equation of the prebiotic chemoton 378

IX. Prebiotic Evolution 383
Introduction 383
Beginnings of Prebiotic Evolution 384
Conditions of biogenesis on Earth 384
Spontaneous genesis of proliferating systems 388
Spontaneous genesis of chemotons 389
External conditions of early evolution 391
The early evolution of chemotons 394
Appearance of Enzyme Activity 396
The problem of primitive catalysis 396
Evolution of RNAs 400
Development of enzyme RNAs 403
The Origin of Coenzymes 409
General considerations 409
Coenzymes are the remnants of eRNAs 410
The Origin of Genes and the Code 414
The origin of genes 414
The origin of RNA polymerases 416
The appearance of DNA 419
The origin of code 422
Summary 424

X. Exobiological Outlook 427
Introductory Remarks 427

Possibility of the Existence of Non-terrestrial Types of Life........ 429
Non-carbon-based life.. 429
Carbon-based life with a non-terrestrial type of metabolism 434
Life of non-terrestrial origin with a terrestrial metabolism... 436
Exobiological Studies in the Solar System.................................. 437
Dry celestial bodies.. 437
Mars... 440
Galilean moons.. 449
Giant planets... 452

XI. Outlines of a Quantitative Biology 453
Introduction... 453
Some Genetic Considerations.. 455
Self-(re)production without genetic subsystem 455
Reproduction by a "technological recipe" 458
Hereditary properties acquired by adaptation...................... 461
General structure of the genetic material............................. 464
Quantitative Treatment of a Prokaryotic Cell.............................. 470
Possibilities and limitations 470
Stoichiometry of the minimum model of prokaryotes 472
The Stoichiometric Principles of Evolution 479
Csányi's general theory of evolution 479
Stoichiometric principles of the evolution of prokaryotes 483
Evolution to eukaryotes.. 486
Evolution of eukaryotes.. 491
Applications in Biotechnology 493
Limits of the prospective development of biotechnology 493
One-step chemical syntheses 496
Chemical syntheses by reaction networks........................... 498
Synthesis of Living Systems.. 500
The objective of synthesis 500
Neobiogenesis.. 502
Possibility of an exact proof.. 506

References 509
Index 539

VI. About the World of Living in General

What is Life?

The ambiguity of the concept of life

The open aim of this book is the creation of the principles for an exact theoretical biology (for such would enable us both to deduce qualitatively and quantitatively the very complex and manifold phenomena of the living world from the basic phenomena of biology, and to project qualitatively and quantitatively the biological phenomena and processes). For this process, experimental data are provided by the systematic arrangement of the most general phenomena of the world of the living; a theoretical basis is created by the chemoton model derived in the first volume of this book, whereas the basis of quantitative treatment is given by cycle stoichiometry (also described in the first volume).

The central concept of biology is the concept of life, as it is qualified for discussing the phenomena and properties of living systems. It is almost unbelievable, but in spite of the tremendous results achieved in biology, the concept of life is as yet not completely clear. It depends on the current "fashionable" research directions rather than on a comprehensive, thorough consideration of which among the multitude of properties of living systems is considered as important by anyone at any time.

Let me present three examples of the unqualified life concept. The first example is the problem of the beginning and end of human life, which has become a serious social and legal question. The end of human life, i.e., the onset of death, is regarded as the time of permanent cessation of the electrical activity of the brain.

That is, this special electrical activity is considered as the single characteristic of the living state. However, the beginning of human life is not associated with to the presence of a characteristic electrical activity, though the determination of the beginning of life also presents a serious legal problem; the problem of murder. The destruction of an embryo that is a few days or months old is not considered murder in the majority of societies (for this purpose, even medicines are officially sold). The destruction of an embryo that is a few months old is allowed in several countries (restricted only by hygienic or sociopolitical reasons); whereas in other countries it is forbidden, but is not considered murder even in the majority of these.

Human beings are the most complex systems of the living world: the determination of the living state is, however, no less problematic at the level of the simplest systems either. An example of this – and this will be our second example – is the question of the living or nonliving state of viruses. This problem has already been dealt with at the beginning of Chapter IV. When Ivanovskii in 1892 discovered the cause of the mosaic disease of tobacco, he thought to have found this cause in a (poisonous) material, and therefore gave it the name "virus" (which means poison). It turned out later that the amount of the virus multiplies in an infected organism. At that time this seemed to be proliferation, and thus it became obvious that the virus is a living being. Later on, Stanley succeeded in crystallizing the virus of tobacco mosaic. Crystallization is, however, a speciality of nonliving materials, thus the living state of viruses again became questionable. However, at this time, viruses were referred to at schools and universities as being the smallest infectious living beings; consequently they became known as such. As bacteria were classified as fungi, and fungi belong to the flora, viruses were taxonomically considered as plants: they were classified as the first phylum of plants by the name Virales. There were viruses of plants and animals, with bacteriophages and rickettsia among them. The belief in the living state of viruses has been strengthened by the discovery that viruses can suffer mutation, i.e., they carry genetic information.

However, when the structure of viruses became known, it was proved that they are not living beings, as – at least in the case of simpler viruses – they consist of a single nucleic acid molecule surrounded by a protein coat. This means that they have no metabolism; no processes take place within them at all that could be called "life." Though their nucleic acid carries special genetic information, they have no reading apparatus that could read this information and carry out the commands contained in it. It is the host cell that is capable of reading the information and carrying out the commands. Thus the virus itself does not proliferate either – only the host cell produces the nucleic acids and proteins of the virus according to the information of the virus nucleic acid that becomes incorporated in it. The formation of the virus from the nucleic acid and protein produced is an agglomeration process that takes place as a result of the effect of purely physicochemical forces. (Later on, the simplest possible "pathogens," viroids were discovered that do not contain even protein. The viroid is just a single, not even very long, RNA molecule.)

The third exàmple is the so-called "square bacterium." A. E. Walsby found microscopic formations of square and rectangular forms when studying the salt precipitates on the surface of the hypersaline pools of the Sinai peninsula that contained vacuolumlike gaseous inclusions. Electronmicroscopic studies showed that these plates have dimensions of several tens of square micrometers, a thickness of 0.2–0.5 μm, and are built of particles ordered in regular hexagonal crystal lattices. Walsby supposed that these formations are bacteria belonging to Archaebacteria. Thus he tried to cultivate them under laboratory conditions, but he did not succeed. He does not report on any phenomenon on life in his paper. In spite of this, he describes them as new, unknown bacteria, i.e., living beings based only on the fact that these square (sometimes triangular or trapezoid) plates contain vacuolumlike gaseous inclusions! The paper was published by the undoubtedly most distinguished scientific journal of the world: *Nature* (Walsby, 1980), and not one biologist raised any objection!

Note that with respect to the beginning of life, the criterion of life was a not accurately defined ontogenetic state of maturity, and its end the presence of a special electrical activity; in the case of viruses, at first the presence of a simple form of proliferation, then that of genetic information; and in the case of square bacteria the only life criterion present was the supposed vacuola!

These randomly taken three examples show clearly that Rizhkov was right. He said in 1958 at a conference in Moscow, which dealt with the philosophical aspects of natural sciences and whose main subject was the nature and origin of life:

> We have seen in the discussion on the origin of life how little we deal with the logic of biology. There was complete confusion when discussing the criteria of life. Some authors said that it was not that important any more to be able to define precisely what we called living and nonliving beings. According to Stanley's opinion it is a question of taste whether we consider viruses as living or nonliving beings. What is this, if not a total failure of the logical apparatus of modern biology? (Rizhkov, 1962)

Views concerning the essence of life

In the course of the history of science numerous views have been expressed about the essence and nature of life. Their enumeration cannot be a task of this book, so the reader is referred to other works (Agar, 1951; Bertalanffy, 1952; Blandino, 1969; Le Dantec, 1900, 1910; Engels, 1963; Gánti, 1971, 1972, 1977, 1979b; Hall, 1969; Hartmann, 1925; Jacob, 1970; Kahane, 1962; Lwoff, 1962; Marquand, 1968; Miller, 1965a, b, c; Monod, 1970; Oparin, 1957; Sándor, 1979; etc.). However, the number of works dealing thoroughly with the nature of life as research work is surprisingly small. The majority of statements in the literature are rather one-sided; they reflect only the specific aspects of a given historical period or research field. For example, overemphasizing the importance of lymph in living beings resulted in the humoral pathology of Hippocrates; motion, which brought animal life into the foreground, was considered by Leeuwenhoek as the only criterion for telling living and nonliving beings apart when performing his microscopic studies; the specific substances of living beings – the organic compounds – came to the fore, and it was thought that they could not be synthesized in the laboratory, but only by living beings by means of "vis vitalis," until, that is, the discovery of Wöhler; the morphological structure of cells became a mysterious "life-carrying" factor (after the development of microtechnique) etc.

All these views were based on true observations; they seemed to be or really were generally valid at a certain time, but they overemphasized one or another of the properties of the living world. In addition to them, however, there are some general properties of living systems, the effects of which are very significant for any outlook upon life. They are therefore worth dealing with in a few sentences.

Metabolism

Plato has written: "... every living being is the same one and identical ...though there is nothing in it which remains the same; the same name is given to it though it is always born anew, and something always perishes in it, in its flesh, in its bones, in its blood and in its whole body." This sentence can be considered as an early formulation of metabolism. It is interesting to note how early the constant renewal of the substance of body is observed, though exact proof for this has only been found by the isotopic tracer method of Hevesi.

Metabolism gained a more concrete meaning in the nineteenth century, partly because the great diversity of compounds constituting living material became known and partly – and this is very important – because Berzelius discovered that metabolic processes occurring in living systems consisted of thousands of catalytic processes, making possible the proceeding of chemical reactions within biological temperature ranges.

In the nineteenth century it also became clear that the chemical processes of metabolism obey the same thermodynamic laws as physical, work-performing processes and that life functions are made possible by the liberation of chemical energy contained in nutrients, and by the utilization of this energy in metabolism. Toward the end of the nineteenth century the basic role of metabolism in life processes had been fully clarified, e.g., Engels defined living beings as metabolizing special substances (protein bodies; Engels, 1963). However, in the second half of the twentieth century the role of metabolism has been pushed back behind information operations.

Proteins

In the organic chemistry of the nineteenth century, proteins, a special kind of compound, became the focus of attention. The concept of protein – and this should be emphasized – was not defined chemically at this time, and quite different substances were meant by it than today: nitrogen-containing, structure-forming, or jellylike organic materials. For example, Engels, who defined life as the mode of existence of proteins, described what he meant by proteins in detail:

> ...the chemical formula of protein is not determined so far, we do not even know how many chemically different proteins exist and ... the fact that the fully unstructured protein is capable of every essential life function such as digestion, elimination, motion, contraction, reaction to stimuli and proliferation has been known only for about ten years. (Engels, 1963)

The present concept of protein was worked out by Fischer and Hofmeister, between 1902 and 1906. According to their findings, proteins are essentially polypeptide chains formed from amino acids with peptide bonds. These chains have none of the properties listed by Engels (although they take an active part in every process of living systems). The protein concept of Engels may be identified rather with that called protoplasm, including nucleic acids. For example, Liebig wrote that it is acknowledged and proven by experiments that plants produce protein-containing compounds and that these compounds are produced from the oxygen of air and water by "vis vitalis" in order to create the tissues and organs of animals (Liebig, 1842).

However, in the same work Liebig characterized "vis vitalis" with such abilities that it is capable of decomposing nutrients, rearranging the elements of nutrients (thereby producing new compounds), forcing the new compounds to combine into new formations, etc. It turned out only five to six decades later, with the elucidation of fermentation, that these properties attributed to vis vitalis are the properties of ferments, or as they are known today, of enzyme proteins. Nevertheless, the secret of the operation and formation of enzyme proteins remained hidden for a further half century. Thus – though the mystery of vis vitalis is banished from organic chemistry – the mystery itself, hidden already behind the protein concept, remained.

This may be characterized well by the words of Claude Bernard:

> In other words: the chemist in his laboratory and the living organism in its apparatus work similarly, but both with their own special method. The chemist may produce the material of living beings, though he will never possess its means, as they originate directly from the morphological structure of living organisms which is beyond the nature of chemistry; and in this sense it is not more hopeful for a chemist to synthesize even the simplest enzyme than to build up the whole living system. (Bernard, 1878, p. 227)

Apparently, for Liebig, Engels, and Claude Bernard the proteins, ferments, became the carrier of vis vitalis, which reflects the mentality of the nineteenth century. It is no wonder that in the first part of the twentieth century biology was dominated by the attitude that the secret of life is hidden in proteins, and that if someone wants to solve the secret of the origin of life, he should study the abiotic genesis of proteins. This is why the paper by Stanley Miller published in 1953 in which he reports on the abiotic formation of organic compounds became a world sensation. There were, actually, earlier works on the abiotic formation of organic material, but they did not receive much attention in the scientific world. However, Miller proved also the abiotic formation of amino acids, which are the building blocks of proteins. It was appreciated that the research into biogenesis obtained thereby a real scientific basis. It was "only" necessary to prove that proteins could also be synthesized from these amino acids without any cooperation of living beings.

The research into the origin of life suddenly gained momentum. Every researcher studied the abiotic formation of amino acids and proteins, and the majority of them reported on the subject as research on the origin of life. Several thousands of papers have dealt with this subject or have reported on such results in the second part of the twentieth century. However, no essential breakthrough has been made by these works concerning the research into the origin of life itself.

Deoxyribonucleic acid

During the 1960s nucleic acids emerged as a serious competitor of proteins as the carriers of "vis vitalis", as it was revealed that the specific properties of proteins are a result of their amino-acid sequence (and to the specific spatial structure owing to this) and that this sequence is determined by the nucleotide sequence of nucleic acids. As nucleic acids – with the aid of proteins – are capable of replication, whereas proteins themselves are not, pieces of information concerning the specific properties of proteins can be inherited only through nucleic acids. Genetics has unambiguously stated that the only carriers of the hereditary properties of living systems are nucleic acids, and in cellular living systems exclusively DNA.

We prove later on that this statement – in such an extreme formulation – is not true, though the majority of properties are in fact inherited through DNA. However, the trouble is that the current attitude in genetics identifies more and more the property with the information concerning it, i.e., the property with the gene, though this is not explicitly stated in the literature of genetics. As the whole of the properties of a living system is the living system itself, and according to a frequently used formulation genes or nucleic acids are the "carriers of properties," it follows that the whole of genes is the living system itself. As the total of genes for an organism is identical with the DNA of the cell, the conclusion is drawn that life is DNA itself. This formulation may be often found nowadays in an implicit form, but it may also be heard explicitly in professional circles.

From this viewpoint it would directly follow that viruses (and also virions) are living systems and that the problem of the origin of life is identical with that of the abiogenic formation of nucleic acids. In fact, in the middle of the 1960s – in addition to research concerning the abiogenic formation of proteins – an ever-increasing emphasis was attached to the spontaneous, abiogenic formation of nucleic acids. These studies, as we discuss later in more detail, have in fact brought a number of usable results important from the viewpoint of the process of the genesis of life.

Lectures delivered nowadays at conferences dealing with the origin of life may be classified into four groups:

1. Formation of small organic molecules under abiotic circumstances;
2. Formation of protenoids (compounds of proteinic character) under abiotic conditions;
3. Spontaneous formation and abiotic replication of nucleic acids;

4. Spontaneous formation of microscopically small structures and microspheres.

Several dozens of conferences have been organized on the "origin of life" during the last decades, and several dozens of books have been published under this title or similar ones. Although neither of them has gone beyond the previously listed four topics, they investigate only the abiotic genesis of the raw materials needed for the formation of living systems rather than the formation process of living systems themselves. It is true that at conferences or in publications dealing with biogenesis sometimes a fifth topic arises: the origin of the genetic code. However, these studies have brought no other results than different and often contradictory speculations. The reason for this is discussed later.

The highlighting of nucleic acids brought an old logical problem to the surface: the question of the priority of the hen or the egg. Evolution theory has unequivocally proved that this is only a pseudo problem. The question is not justified as it projects the present state back into the geological past, not taking into account biological evolution – the continuous process of development on a geological time scale.

The problem of the hen and the egg took a new form in the past century: if organic substance can be produced (in nature) only by living systems, but for the formation of living systems organic substances are needed, which one preceded the other? Research has now given an unambiguous answer to this question: organic substances are formed in nature everywhere without the cooperation of living systems provided the conditions for this are ensured. Thus this question was raised earlier as a consequence of some lack in our knowledge.

Nevertheless, the problem of the hen and the egg was raised again recently in a new form, and this is why the excellent work of François Jacob entitled "La logique du vivant" obtained in a Hungarian translation the title: "The hen and the egg." Molecular biology has unequivocally clarified the fact that the information on the sequence of amino acids determining the structure of the many thousands of proteins occurring in living organisms is coded in the nucleotide sequence of DNA of cells; thus such special proteins catalyzing the life processes cannot be formed without the information enclosed in the nucleic acids. Moreover, molecular biology has also established that the replication of nucleic acids in cells takes place by means of complicated processes catalyzed by complex enzyme proteins, the information about which is contained in nucleic acids. Thus proteins cannot be produced without nucleic acids, and neither can nucleic acids be produced without proteins.

The question is: which one existed earlier, the nucleic acid or the protein? Today this is the main question concerning research into the origin of life. This, however, is a typical hen-and-egg-type question. Later on we show that the process of the genesis and evolution of life can be treated without the need to raise this question.

The standpoints of some outstanding philosophers

It was mentioned that numerous researchers expressed their opinions and views about the nature of life in the course of the history of science, but there were only a few who would have put this question into the focus of their studies, or at least would have done general, comprehensive research about life itself. Among those who did, the views of four scientists on the nature of life are worth mentioning: those of Leibniz, Ervin Bauer, Schrödinger, and Bertalanffy. They are mentioned here and parts of their work are cited in order to show that the same conclusions can be drawn on the nature of living systems in different ages, at different levels of science, with different education and attitude and by the application of different languages and conceptual apparatuses, and that these conclusions are in accordance with principles following from chemoton theory.

Gottfried Wilhelm Leibniz

This giant of the history of ideas who dealt with everything from history and politics to alchemy and, casually, also discovered differential calculus wanted to create a uniform, closed system of philosophy including the whole of the world. As part of this system, he had to deal with living beings in such a way that his statements would be in accordance not only with the biological properties of living systems but with his comprehensive system of the world (and with his own religious beliefs) as well. Presumably, this is why he succeeded in making generalizations on living systems that are worth considering even after nearly three centuries, in spite of the backwardness of the biology of his time. We may best get acquainted with his views about living systems from his short essays and letters in which he is arguing with his critics or explaining his opinion.

First, Leibniz thought living systems to consist of body and soul. It is especially interesting to consider what he meant by "soul":

> ...organic bodies are never without a soul, and souls are never fully deprived of organic body, although it is also true that there is no material of which it could be said that it is bound

always to the same soul. Thus I do not allow for fully separated souls in a natural manner, neither for created spirits devoid of body, in which views I agree with the opinion of many old Fathers of the Church. (Leibniz, 1705)

Several years later he felt the need to expose his view about the soul in more detail:

... I attribute the acting basic principle not to the mere or first material which is passive and has only resistance and extension, but to the body, that is to the endowed or second material which, in addition, contains also first entelechia, i.e., basic acting principle. ...the resistance of mere matter is not acting, only suffering. It has, namely, resistance or impenetrability by which it resists penetration, but it cannot push back the thing which wants to penetrate it, unless it has elastic force as well, which has to be derived from the active force of the matter originating from motion. ...this basic acting principle, this first entelechia is a real life principle (principium vitale) which has a perceiving ability as well, and which is imperishable. And this is just what I consider as the soul of animals.

Let us not forget, however, that in my opinion every soul, every spirit, every entelechia is not only imperishable, but to every entelechia or every basic life principle a natural machine is also connected which we call organic body, though on the whole, this machine constantly changes, it maintains its shape and is always restored, as the ship of Theseus. Thus we may be sure that a small part of the matter which we obtained at our birth remains in our body, although this machine transforms, grows, diminishes, gets wrapped up and develops repeatedly. Therefore not only soul is perpetual, but something always remains also of the animal, though a certain animal must not be declared as immortal as the shape of the animal is not maintained: just as the worm and the butterfly are not the same animal, though they have the same soul. (Leibniz, 1710)

Several years later, in "Monadology," he summarizes his views as follows:

63. A body belonging to a monas which is its entelechia or soul, comprises together with the entelechia what we call a living thing, and together with the soul what we call animal. This body of the living or the animal is always organic, as every monas reflects in its own manner the universe and as the universe is controlled by an ultimate order, it is necessary that a certain order should also be present in the representatives be present, i.e. in the ideas of soul and, consequently, also in the body according to which soul represents the universe.

64. Thus the organic body of a living thing is a sort of divine machine or natural automaton which infinitely surpasses artificial automata. The machine designed by human art is not a machine in its every detail, and for example the cog of a brass wheel has such details and parts which do not have anything artificial any more and do not have anything to reveal the function of the machine. In contrast, the machines of nature, i.e. living beings, are machines up to their smallest details, to infinity. This is the difference between nature and art, or more correctly, between divine and human art. (Leibniz, 1714)

On rephrasing all this to our present language it means that living beings are natural automata differing, however, in their organization and operating principles from automata in the technological sense. They are not machines in the mechanical sense, but – as he writes somewhere else – machines that do not originate from themselves, i.e., those that are derived from other natural machines so that they carry pieces of them and through this, their soul. They can do this as, unlike machines in technology, their constituents are also automata, and so their small parts also carry the basic acting principle (today we would rather say an organization principle) ensuring the internal order of these natural machines. In contrast to machines of technology, the functioning of living systems as natural machines is ensured not by an external designer but by an inner basic acting principle that he calls the soul of the living system.

Ervin Bauer

The name and activity of this outstanding biologist of Hungarian origin who met with a tragic fate as a victim of the personal cult in the Soviet Union is not recognized enough in the international scientific literature. We can justly call him one of the first theoretical biologists. The concept of life and the most general laws of life functions represented the center of his life's work. He published two books – the first appeared in German in 1920, the second in Russian in 1935. The same theoretical subject is dealt with in both, only at different levels of development. These works of his appeared also in Hungarian in 1963 and in English in 1982 (Bauer, 1963, 1982).

Bauer wanted to deal with theoretical biology consciously and openly, and therefore he defined very carefully the subject and aim of theoretical biology:

... If we attribute to living, organized matter own laws of motion which are valid for living matter and only for living matter, i.e. they form its attributes, mode of existence, then these laws have to be valid in every outward form of living matter...

... The theoretical treatment of a phenomenon consists of abstracting the generally valid, the regular from the multitude of specific, concrete phenomena, and of scrutinizing and analyzing this specific derived regularity and presenting its possible forms.

If we succeed in abstracting the general laws of motion corresponding to a given material state and in deriving the individual empirical laws as manifestations of these general laws of motion under given conditions, only then can we speak about a theoretical and general science of the field in question. The collecting and the discussion of individual laws obtained by generalization is the task of descriptive and experimental science in every branch of science.

Evolution theory needs deepening in order to be the historical theory of living matter. This step can only be made in the knowledge of the general laws of motion of living matter. Consequently, the next task of theoretical biology is the elaboration of the general laws of motion for living matter, i.e. that of its theory. The task of theoretical biology is to derive the various empirical laws and data of descriptive and experimental biology as special cases of laws and ideas concerning living matter which are more general and go beyond direct experience. These general laws should be characteristic for living matter, i.e. the specific laws of motion of living matter, at the same time, should be valid for the mode of existence and laws of motion of living matter always and everywhere, in whatever special form of it. Thus these laws should be valid for every living matter, and only for living matter.

After this, Bauer expounds the subjective of his own work:

... On the one hand, we want to characterize material systems called living systems on the basis of their particular properties separating them from other systems, and on the other hand, we cannot name even a single property without great difficulties, which would characterize these systems and simultaneously would separate them from other systems. In fact, it is interesting, how unambiguously and confidently is the expression 'living' used in science and in the standard language, considering the difficulty and uncertainty as well as the truthfulness with which science tries to define this concept.

In the present work life is tried to be defined so that the correctness of this definition could be checked, and this checking is also given a try. Thus our task would be to comprise and express the features characteristic for every living being and only for living beings in one or several laws.

Finally, Bauer formulates the principle of life in a single basic law:

Living systems and only living systems are never in equilibrium; they do work continuously at the expense of their free energy content against the establishment of the equilibrium which should exist under the given external conditions according to the laws of physics and chemistry. ... Such general laws, if they do not contradict facts, forming thus the basis for all studies, are called basic principles. For example, in mechanics such principles are the 'd'Alambert principle,' the 'Hamilton principle,' the 'principle of inertia,' etc.

Let us call our principle the "principle of constant non-equilibrium" of living systems. Bauer provides this principle with a mathematical formulation:

The free energy of an isothermally closed living system is smaller in equilibrium than that of a *nonliving system if the sum of the potential differences measured in absolute units has been initially identical in them.* Under such conditions, the free energy of the living system is smaller in equilibrium than that of the non-living system just by the work done by the living system against the equilibration process.

Thus if the free energy of the living system is in equilibrium equal to F, the sum of the work factors – differences in pressure, concentration, electric potential, etc. – is equal to the changes caused by them in a time interval of t is equal to X, the corresponding data for nonliving systems are denoted by F′, x′ and Δx′, then:

$$F' - F = \sum_{i=0}^{n} X_i \frac{\Delta x_i}{\Delta t} \Delta t - \sum_{i=0}^{n} X_i' \frac{\Delta_i'}{\Delta t} \Delta_t \qquad [6.1]$$

... On decreasing gradually the time intervals, the limiting value of the sum is obtained which provides the real value of work for infinitely short time intervals:

$$F' - F = \int_{t=0}^{t=\infty} X \frac{dx}{dt} dt - \int_{t=0}^{t=\infty} X' \frac{dx}{dt} dt \qquad [6.2]$$

This is the mathematical basis of our quantitative biological basic principle.

Behind this ponderous formulation there is a simple basic principle formulated again by Bauer as follows: "Whereas in non-living systems the processes of equilibration are determined by the direct properties of the structure, the situation in living systems is different, as in the course of equilibration the properties of the structure also change."

The energy state of all operating systems is higher than that of non-operating ones, i.e., in their equilibrium: one of the basic principles of irreversible thermodynamics is that a system may be kept in a state far from equilibrium only by a constant energy consumption. This concerns of course both living and nonliving systems. This energy is provided for in both living and nonliving systems by an external energy source. Thus, in this respect, there is no difference between them. If we stop the flow of energy from this external source into the system, the nonliving system approaches its equilibrium state by releasing its energy. Living systems, however, do not behave this way. They make the system function by the decomposition of their own parts and by using the energy thus obtained further than it would be possible as a result of the "inertia provided by their own speed of operation." This is analogous to a steam engine beginning to burn a part of its own screws and wheels after having used up the coal, ensuring thereby the running of the engine as if it were still using the energy of the coal.

This statement of Bauer casts light on a basic property of living systems. His tragedy was that he was far ahead of his time, as at the time of his activity the basic principles of irreversible thermodynamics elaborated by Onsager, Prigogine, and Gyarmati were not born (Onsager, 1931; Prigogine, 1947; Gyarmati, 1960, 1970). This theory makes it possible to approach living systems from the side of their operation. Neither had molecular biology been born, with the aid of which the mechanism of this difference could be explained.

Erwin Schrödinger

Schrödinger was already a successful, distinguished physicist when he began to deal with the essence of life. With his unbelievable farsightedness he discovered the two most important and most general properties of living systems in a short time and in such a way that their formulations remained valid even after four decades. His considerations on life have been published in a booklet (Schrödinger, 1944). The main topics in genetics at this time were mutations and the study of chromosomes. Schrödinger studied systematically the newest results of genetics of his time and his attention was caught especially by the regularities of mutations. He wrote:

> It reminds a physicist of quantum theory – no intermediate energies occurring between two neighbouring energy levels. He would be inclined to call de Vries's mutation theory, figuratively, the quantum theory of biology. We shall see later that this is much more than figurative. The mutations are actually due to quantum jumps in the gene molecule.

At that time it was already known by geneticists that the properties somehow hidden in genes can be found in chromosomes, and they had guessed that the gene is probably a very specific molecule. But what kind of molecule was it? Schrödinger calculated that upon dividing the volume of chromosomes by the number of genes contained in them such a small volume is obtained that only a few atoms could find space in it. This number of atoms was too small; consequently it was not possible that the properties were coded by a mechanism obeying statistical laws. The code hidden in genes had to be deterministically defined. This assumption seemed to be supported by the fact that in mutations caused by irradiation "only one single event takes place" as concluded from the dose–effect relations, i.e., a variation in one atom or in a small group of atoms brings about the change in a whole property.

As the structural element of this deterministically defined "gene molecule," he introduced the concept of the "aperiodic solid body":

> A small molecule might be called a 'the germ of solid.' Starting from such a small solid germ, there seem to be two different ways of building up larger and larger associations. One is the comparatively dull way of repeating the same structure in three directions again and again. That is the way followed in a growing crystal. Once the periodicity is established, there is no definite limit to the size of the aggregate. The other way is that of building up a more and more extended aggregate without the dull device of repetition. That is the case of the more and more complicated organic molecule in which every atom, and every group of atoms, plays an individual role, not entirely equivalent to that of many others (as is the case in a periodic structure). We might quite properly call that an aperiodic crystal or solid and express our hypothesis by saying: We believe a gene – or perhaps the whole chromosome fibre – to be an aperiodic solid.

Watson and Crick solved the problem of the structure of DNA nine years later. It has been proved that the DNA molecule (and thus the gene) is really of an "aperiodic crystal" structure (Watson and Crick, 1953a), as is the chromatid.

Schrödinger also realized that this structure is capable of storing information:

It has often been asked how this tiny speck of material, the nucleus of the fertilized egg could contain an elaborate code-script involving all the future development of the organism? A well-ordered association of atoms, endowed with sufficient resistance to keep its order permanently, appears to be the only conceivable material structure, that offers a variety of possible ('isomeric') arrangements, sufficiently large to embody a complicated system of 'determinations' within a small spatial boundary. Indeed, the number of atoms in such a structure need not be very large to produce an almost unlimited number of possible arrangements. For illustration, think of the Morse code...

What we wish to illustrate is simply that with the molecular picture of the gene it is no longer inconceivable that the miniature code should precisely correspond with a highly complicated and specified plan of development and should somehow contain the means to put it into operation.

Nine years later, Watson and Crick proposed a mechanism for the storage of hereditary information corresponding to this, which was eventually proved to be correct (Watson and Crick, 1953b).

However, Schrödinger thought even deeper than to be satisfied with this answer to the question, "What is life?" He saw clearly that however important and indispensable is the coded information for living systems, life itself is not identical with the information about it. He wrote about this the following:

What is the characteristic feature of life? When is a piece of matter said to be alive? When it goes on 'doing something,' moving, exchanging material with its environment, and so forth, and that for a much longer period than we would expect an inanimate piece of matter to 'keep going' under similar circumstances. When a system that is not alive is isolated or placed in a uniform environment, all motion usually comes to a standstill very soon as a result of various kinds of friction: differences of electric or chemical potential are equalized, substances which tend to form a chemical compound do so, and temperature becomes uniform by heat conduction. After that the whole system fades away into a dead, inert lump of matter. A permanent state is reached, in which no observable events occur. The physicist calls this the state of thermodynamical equilibrium or of 'maximum entropy.' Practically, a state of this kind is usually reached very rapidly...

... It is by avoiding the rapid decay into the inert state of 'equilibrium,' that an organism appears so enigmatic.

Let us notice that these thoughts are in accordance with those of Bauer of several decades earlier and even with those of Leibniz from three and a half centuries ago! Each of them broke through the superficial phenomena and saw the essence, the basic principles. Finally, Schrödinger also used a thermodynamic formulation, only he utilized the concept of entropy instead of that of free energy, but in this respect both may be used equally well:

Every process, event, happening – call it what you will; in a word, everything that is going on in Nature means an increase of the entropy of the part of the world where it is going on. Thus a living organism continually increases its entropy – or, as you may say, produces positive entropy -- and thus tends to approach the dangerous state of maximum entropy, which is death. It can only keep aloof from it, i.e. alive, by continually drawing from its environment negative entropy – which is something very positive as we shall immediately see. What an organism feeds upon is negative entropy. Or to put it less paradoxically, the essential thing in metabolism is that the organism succeeds in freeing itself from all the entropy it cannot help producing while alive.

It is also worth citing how Schrödinger summarized his views about life:

The unfolding of events in the life cycle of an organism exhibits an admirable regularity and orderliness, unrivalled by anything we meet with in inanimate matter. We find it controlled by a supremely well-ordered group of atoms, which represent only a very small fraction of the sum total in every cell. Moreover, from the view we have formed of the mechanism of mutation we conclude that the dislocation of just a few atoms within the group of 'governing atoms' of the germ cell sufficient to bring about a well-defined change in the large-scale hereditary characteristics of the organism.

An organism's astonishing gift of concentrating a 'stream of order' on itself and thus escaping the decay into atomic chaos – of 'drinking orderliness' from a suitable environment – seems to be connected with the presence of the 'aperiodic solids,' the chromosome molecules ...

Ludwig von Bertalanffy

The beginning of his activity coincided with that of Erwin Bauer, and the majority of his works were published in the 1920s and 1930s. Later, he summarized his views on life in a more advanced form in a book entitled "Problems of life" (Bertalanffy, 1952). He considered living beings as systems having internal order:

The phenomena of life – metabolism, irritability, reproduction, development, etc. – are found exclusively in natural bodies which are circumscribed in space and time, and show a more or less complicated structure, bodies which we call 'organisms'. Every organism represents a *system* by which term we mean a complex of elements in mutual interaction.

... Thus the characteristics of life are characteristics of a system arising from, and associated with the organization of materials and processes ...

... The problem of life is the problem of organisation ...

It is also worth noting that according to Bertalanffy organization important not only at a morphological level (indeed, first of all not at that level) but also for processes. At the time of his work biochemistry had already made enough progress so that he was able to recognize that the processes are mainly chemical processes; therefore organization should be sought primarily at a chemical level, and only after that can it be continued at the level of the cell, then again even later at that of organisms:

Similar considerations apply to the processes of life. So long as we consider the individual chemical reactions that take place in a living organism we are unable to indicate any basic difference between them and those that go on in inanimate things or in a decaying corpse. But a fundamental contrast is found when we consider not single processes but their totality within an organism or a partial system of it, such as a cell or an organ.

The main task of biology therefore is to establish the laws governing order and organisation within the living. Moreover, as we shall see presently, these laws are to be investigated at all levels of biological organisation – at the physicochemical level, at the level of cell and of the multicellular organisation and finally at the level of communities consisting of many individual organisms.

Bertalanffy considered viruses as some "elementary biological units" (because at that time they were believed to be reproducing), but as we see later in connection with his definition of life, he did not consider them as living. For him, the basic unit of life was the cell:

Thus the elementary biological units show growth and co-variant reduplication, but the complete complex of synthetic action, namely of building up organic molecules *de novo,* remains a privilege of the cell as a whole.

So we come to the cell as the fundamental unit of life. The cell, i.e. a unit whose essential components are the nucleus and the cytoplasm, is the simplest system known that is capable of autonomous life. It is a quite astonishing fact that all living beings from a minute unicellular alga up to a thousand-year old giant tree, and from the amoeba to man, are formed by variation and multifarious combination of this one building unit. This fact indicates the presence of a basic structural law.

It should inevitably be mentioned here that at the time of the elaboration of Bertalanffy's theory the construction and organization of prokaryotic cells was not known, thus he could not know that there is no nucleus in prokaryotic cells in the morphological sense. However, based on his concept it is certain that he would have considered the prokaryotic cells as living even if he had our present knowledge. Because of this lack of appropriate knowledge he could not solve the problem with which he deals most in his book: the accurate separation of organization levels – the hierarchic order of the structure of the living world.

At the end of his book he comes back to the basic principles:

Living forms are not in being, they are *happening*: they are the expression of a perpetual stream of matter and energy which passes the organism and at the same time constitutes it. ...living organism is ... an open system ... in a steady state.

That is, from the point of view of life, not the living forms, but the happenings occurring in them are important, in the course of which matter and energy stream through them and depart from the organism while also constituting it. The living organism is an open system being in a steady state.

This last statement of Bertalanffy became a platitude in biology. The second part of this statement, namely, that the living system is in a steady state – though it is used often – gave rise to many misunderstandings, because, unfortunately, it is not true.

Finally, Bertalanffy attempts to give a general definition of life, which, however, he himself does not feel satisfying: "A living organism is a hierarchical order of open systems which maintains itself in the exchange of components by virtue of its system conditions." However, he notices that "viruses are outside our definition, because they are not capable of reproduction outside the cells and thus they do not dispose of all conditions of autonomous self-preservation through metabolism."

The Most General Properties of Living Beings

Empirical facts providing the basis for the theory

In what follows, some of the most general properties of the living world will be briefly discussed. More exactly: as the features discussed so far are not suitable for deciding unambiguously what can be considered a living system and what cannot, in the following, when studying the properties, systems are investigated belonging to the field of biology, independently of whether their being living is doubtful or even may be denied (viruses, viroids). The properties discussed are with several exceptions commonplace for the biologist but certainly not in their formulation and relations. Empirical facts mentioned here form the practical basis of the theory. Therefore it is especially important to think them over thoroughly – not only in themselves, but also in their relation to the conclusions deduced. Although it is obvious, for the sake of what follows we mention here that these experiences originate exclusively from the terrestrial living world.

Thesis-like formulations of the facts to be discussed are as follows:

a. *Every living being contains a significant amount of (fluid) solution.* The major part of this solution is aqueous, but to a small extent, apolar liquids (oil drops, the internal parts of two-dimensional liquid membranes) may also serve as solvents (but only together with the aqueous solution). Exceptions: Viroids, viruses, and dried-up (but viable) living beings do not contain fluid solutions.

b. *Every living being is separated from its surroundings by a limiting wall.* The limiting wall is either a mere two-dimensional liquid membrane, or together with this, a solid wall (or walls) consisting of organic or inorganic molecules, occasionally of both. Exception: Viruses having a protein shell but not having a two-dimensional liquid shell and viroids not having even a protein shell.

c. *Every living being is built of dissolved inorganic ions, organic molecules of small molecular masses, and organic macromolecules according to the following:*

   - ❖ Dissolved inorganic ions. Exceptions: viruses, viroids, and dried or frozen living beings in which the ions are not in dissolved state.
   - ❖ Small organic molecules. Exceptions: viruses, viroids.
   - ❖ Organic macromolecules:
     - Nucleic acids (DNA + RNA). Exceptions: none.
     - Proteins (skeletal and enzyme proteins). Exceptions: viroids.
     - Polymeric hydrocarbons. Exceptions: viruses, viroids (perhaps also myco- and thermoplasms?).

   In addition to the substances listed here there may be other compounds present, but not in general.

d. *The functioning of living beings is based on chemical work.* Exceptions: viruses and viroids, because although they are replicated by chemical processes, these are the chemical processes of the host cell.

   In the vast majority of living beings, in addition to chemical work, mechanical and/or electrical work can also be found, but they too are always based on chemical processes; e.g., chemical work is transformed into mechanical work. The transformation between chemical and electrical work may proceed in both directions.

e. *Processes taking place in living beings are ordered into chemical networks – the so-called metabolic networks – which also mean that in the organism any reaction occurring as a consequence of life functions is related to every other reaction; they are in mutual interaction.* Exceptions: viruses, viroids.

The many thousands of reactions taking place in living beings are thus not an unorganized agglomeration of chemical processes but their well-organized system. This organization is destroyed by death, as new chemical processes of decomposition set in that are not organized in a system. Thus the essential difference between the states before and after death is the difference in the character of the metabolic network.

f. *The basic reaction network of metabolism is the carbohydrate–organic acid network.* This is an especially important fact – perhaps a decisive one from the point of view of the origin of life. It is surprising how common the bases of metabolism are in spite of the extraordinary variety of living beings. In autotrophs, the building in of carbon atoms occurs either in a self-reproducing carbohydrate cycle (Calvin cycle) or in a self-reproducing organic acid cycle (reductive tricarboxylic acid cycle). In heterotrophic organisms the decomposition of nutrients takes place in a carbohydrate–organic acid network. The synthesis of every other compound (amino acids, nucleotides, terpenes, steroids) is built on this network and is often connected to the carbohydrate–organic acid basic network through a linear synthesis path having no stoichiometric feedback.

g. *The functioning of the metabolic network is stoichiometrically regulated and program controlled.* This very important fact is hardly, or at most only indirectly, realized by current chemistry and biochemistry. As we know, stoichiometry calculates only with amounts and not with concentrations, and it deals only with the end results of processes without taking into consideration their time course. A cell has to produce in one cycle the four deoxynucleotides for the synthesis of DNA in such amounts in which they are present in the DNA from the individual nucleotides, and it has to synthesize an exact amount of the material of the cell wall needed for the geometrical coverage of the membrane limiting the cell, independently of time and concentration relations. The whole metabolic network is also stoichiometrically self-regulated in the sense that, from a few (or relatively few) nutrients with the secretion of a few waste materials, it is capable of duplicating its internal system containing several thousands of compounds, i.e., of synthesizing each of its internal compounds in appropriate amounts. In nutrient media of different compositions the stoichiometry of the network varies so that both the constancy of the composition of DNA and the self-reproducing ability of the whole system remain ensured. Thus the composition of DNA controls directly, in its stoichiometry, the stoichiometry of the network as well.

Stoichiometric regulation and control regulate the composition of living systems.

h. *The functioning of the metabolic network is also regulated and program-controlled kinetically.* Kinetics deals with the time course of processes by taking into account the concentration of components and the reaction rate constant (referred to us k values). However, in what follows, by kinetic regulation and control we shall arbitrarily mean, only control through concentration, keeping k constant. Kinetic regulation controls the driving force of the functioning of living systems. Thermodynamics considers chemical affinity as the driving force of chemical reactions:

$$A_i = \sum_{i=1}^{k} \nu_{ij}\phi_j \qquad [6.3]$$

where A_i is the chemical affinity of the i-th reaction, and ϕ_j is the chemical potential of the j-th component. (In thermodynamics, chemical potential is usually denoted by μ; ϕ is used here in order to avoid mixing it up with the stoichiokinetic constant.) Chemical potential is composed of two members:

$$\phi_j = \phi_j^0 + RT\ln c_j \qquad [6.4]$$

where ϕ_j^0 is the standard chemical potential characteristic for the given component, c_j is the concentration of the j-th component, R is the universal gas constant, and T is the absolute temperature.

The stoichiometric constants ν_{ij} of a given reaction, and the values of ϕ_j^0 and R are constant; thus at constant temperature the driving force of a reaction can only be regulated by the variation of the concentrations of the components.

As according to point d. the functioning of living systems is based on chemical work, the resultant affinity of the chemical reactions taking place in it should be regarded as the driving force of the operation of living beings. This resultant affinity is regulated by kinetic control so that in the metabolic network the concentrations of the components are kept at the optimum value through the many feedbacks of the system.

Equations [6.3] and [6.4] have been deduced under classical thermodynamic conditions, thus they are valid for equilibrium or near-equilibrium states and diluted solutions. Living systems are, however, never in equilibrium (only eventually systems capable of life but not living; see later), and neither can their solution be considered diluted. Lengyel and Gyarmati have shown that in the nonlinear ranges of thermodynamic affinities cannot be regarded as the real driving force of chemical reactions (Lengyel and Gyarmati, 1981a, b, 1986).

Although the Gyarmati—Li theory seems capable of deducing exactly the chemical driving force in the nonlinear region, this theory has not yet been published. Nevertheless, it is certain that the driving force can be varied also in this region only by changing the concentrations. Thus, although the relation between concentration and driving force will be different in this case from that is given in Eqs. [6.3] and [6.4], this will not change the fact that in the metabolic network kinetic control and regulation are realized by changes in the concentrations.

i. *The functioning of the metabolic network is catalytically overregulated and over-controlled.* The presence of a catalyst accelerates reactions without, however, changing the driving force of the reaction. Thus the effect of the catalyst may be interpreted as changing the value of the rate constant k, increasing thereby the reaction rate at a constant concentration. In the living world, the rate of life processes is very important in both adaptation to the environment and fights between species and individuals. Every member of the living world contains specific, effective catalysts. Exceptions: some viruses and viroids. These catalysts are, almost without exception, protein enzymes, but recently enzymatically active RNA has also been found (see the origin of enzymes in Chapter VII). Catalysts in living organisms are highly specific and they accelerate the reactions of the metabolic network to a very great extent: about 10^6–10^8 times.

Thus catalytic regulation and control govern the time course of the function of the metabolic network. This is possible because of variations in the amount of the catalyst (and as the volume is usually constant, also in its concentration), in the activity of the catalyst (allosteric inhibition and acceleration, chemical modification, isoenzymes), and also because of competition. Thus catalytic regulation operates in the composition range defined by stoichiometric regulation and at a driving force determined by kinetic regulation, as it were built on these.

j) *The operation of living systems is governed by coded hereditary information stored in the DNA of the organisms.* Exceptions: viruses and viroids that contain hereditary information, but these govern the operation of the host cell. In viroids and some viruses the hereditary information is coded in RNA rather than in DNA.

The hereditary information is coded by the nucleotide sequence of DNA; this sequence, in turn, determines the amino-acid sequence of enzymes (in the case of RNA enzymes, the nucleotide sequence), which determines the catalytic activity and specificity of the enzyme through the formation of the globular structure of the enzyme. Thus DNA fulfills its controlling function first through governing the catalytic regulation, and in fact by varying the k values of individual reactions.

Conclusions

1. *If points* a., d., *and* e. *are true, i.e., the functioning of living beings is based on chemical processes that form a network, ordered in the chemical state space, then living systems are fluid (chemical) automata for which the principles of fluid automata described in the first part of this book should be valid, and the methods for their treatment should be applicable.*

This statement provides an unambiguous theoretical background for numerous findings about living beings that could not be explained so far, and it may lift some apparent contradictions. First it makes understandable the contradiction of the simultaneous existence of constancy and change, which is found in the works of Plato, Leibniz, Cuvier, Goethe, Claude Bernard, and a great number of other scientists and philosophers in some form or other. In fluid (chemical) automata, unlike in mechanical and electrical automata, not only an energy stream passes through the automata, but the mass flow carrying the energy stream itself constructs a significant part of the constituents of the automata. Thus it follows from the basic operating principle of fluid (chemical) automata that, although their organization remains constant during operation, their constituents undergo constant changes.

Thus it becomes possible to interpret the contradiction of "order in disorder" as well as the persisting lack of even a suitable approach to the basic organizational principles of living organisms in the history of science. The problem of inner organization in living things is constantly recurring in scientific history, surfacing either as "vis vitalis" or "basic acting principle" or "spirit," as with Leibniz, or in the machine metaphors of mechanical materialists, or in the "inner order penetrating the mass" (as described by Buffon in his "Histoire naturelle des animaux"). As we have seen, Leibniz regarded living beings as natural machines, or divine automata, whereas for Bertalanffy they represented organized systems, or in current terminology "cybernetic systems." The multitude of formulations and expressions refer essentially to the same thing, namely, to the internal organization of living systems described in terms of the apparently most appropriate concepts of that period. However, the attempt at revealing this organization was fruitless: basically this is organization in the chemical state space, rather than geometrical organization in space.

The possibility of organization in the chemical space was beyond all comprehension so that François Jacob, in his famous book (cited earlier), wrote the following as late as 1970, when the metabolic network maps were already displayed on the walls of biologists' and biochemists' offices:

> ...However, there is no evidence that the transition between organic and living matters will ever be successfully analyzed. Perhaps it cannot even be estimated with what probability living systems appeared on Earth. Genetic code is assumedly universal, because anything which has succeeded so far in living, originated from a single predecessor, and an event occurring only once has no measurable probability. It is to be feared that this subject will fail among the incoherent, not justifiable hypotheses concerning it. It is easily imaginable that the origin of life will become a new focus of abstract debates between schools based on metaphysics instead of scientific foresight. ...No molecule is e.g. capable of reproduction on his own. The ability for self-reproduction appears first in the simplest integron worth calling living: i.e. in the cell. (Jacob, 1970)

To regard living systems as fluid automata makes understandable the sentence of Leibniz cited earlier, namely, "The machines of nature, i.e. living organisms, are machines down to their smallest details, to infinity." This sentence today has to be modified only in that what was considered at the time of Leibniz "infinitely small," the molecular dimensions, can be well studied now. But fluid (chemical) automata – and thus also living systems – are really machines down to their smallest details, even in the microscopic range. According to Leibniz "This makes the difference between nature and art, or more correctly, between divine art and that of ours."

2. *Not every fluid (chemical) automaton is living. Living systems are fluid (chemical) automata of a specific organization.* This statement in itself is not yet a definition of living systems, but it helps us to understand life. It contains, on one hand, the old recognition that living beings possess a specific internal organization and, on the other hand, it says that the laws of this specific organization can be deduced from the laws of fluid automata. This is very important, because as we have shown in the first volume, the organization and operation of fluid (chemical) automata may be described in an exact, scientific way and thus there is some hope that the internal organization, specificity, and operation of living systems can also be described by exact scientific methods.

3. *Life is nothing more than the operation of these specific fluid (chemical) automata. Living is only what operates and when it operates.* Neither is this statement a definition; it serves only the clarification of concepts. "To live" is a verb and as such it means an activity, change, or operation. In spite of this, the literature in biology often regards systems as living in which these specific changes do not take place, e.g., dried spores. Though they are capable of living as - under suitable conditions - processes which can be characterized by the verb "live" may start in them, in their dried state they are not living, but only "capable of living."

If we compare the conclusions 1 and 3, the basis of the principle of "constant non-equilibrium" formulated by Bauer becomes understandable. As we have shown, this principle says that living systems are never in equilibrium, and on their approaching the equilibrium state, excess energy is liberated that delays the attainment of equilibrium; i.e., the living system "fights" against attaining the equilibrium state.

The non-equilibrium state of living systems follows obviously from the fact that only operating systems are living. Equilibrium, namely, means the cessation of functioning. The excess energy loss follows from the statement in conclusion 1 that the "constituents" of the system consist (partly) of the mass flow passing through the system. In equilibrium, namely, the mass flow ceases to exist and, at the same time, the constituents that have been formed dynamically from the mass flow also disappear. (The materials of the internal, closed mass flow, and thus the materials of untapped cycles and networks and the materials of static structures are conserved.)

Thus the principle of "constant non-equilibrium" formulated by Bauer automatically follows from the hypothesis that living systems are operating fluid (chemical) automata. At the same time, it is also obvious that this principle is characteristic of fluid (chemical) automata rather than of the living world. Thus, though this principle should be obeyed by every living system, the principle of "constant non-equilibrium" is not a particular principle of life.

4. *If having systems are specific fluid (chemical) automata the operation of which is life itself and the specificity of which consists of their internal organization, then the basic task of theoretical biology is to investigate which mode(s) of the internal organization of fluid automata result(s) in properties possessed by having systems.* This organizational mode should obviously be sought in the abstract chemical state space and for this purpose the exact methods successfully used for the treatment of fluid chemical machines in the first part of this book should be expediently applied. The main task of this part is to disclose this organizational mode and to outline the origin of life and evolution of the having world on this basis. This aim is in full accordance with the main objective of theoretical biology as formulated by Bauer or Bertalanffy, with the points of view of Leibniz or Schrödinger and with that of all those who guessed the essence of life to be in metabolism, in regulated operation, or in internal organization. On the other hand, it is in contradiction to those who think the essence of life will be found in some chemical substance (organic materials, proteins, DNA) or suppose it to be of a metaphysical nature.

Thus wc look for a system at the level of chemistry, in accordance with the idea of Bertalanffy and others, that shows the basic properties characteristic to having systems during its operation and whose time course is regulated by enzymes and controlled by DNA through enzyme synthesis according to the genetic program. The essence of life is hidden in this system the operation of which the enzyme–nucleic acid complex is capable of regulating and control.

A basic error of current biology is the widespread view that the essence of life is represented by the hereditary information contained in the nucleotide sequence of DNA. However, information in itself does not exist. Information is nothing if there is no system capable of reading, interpreting, and utilizing it. DNA in itself, whatever its sequence, is a chemical substance, a compound, a macromolecule, nothing more. At present, there is no difficulty in the artificial synthesis of any kind of DNA with a predetermined sequence. Therefore today it is obvious that any DNA of whatever sequence is not yet a living system. Consequently, it is basically wrong to study the origin of life by investigating the "spontaneous" genesis of nucleic acids with predetermined sequences. Sequence has an information-carrying meaning only together with the metabolic network, and also only after the appearance of sequence-dependent catalysts: the enzymes. The genetic code, the hereditary program, could be developed only in a co-evolution with the metabolic network; thus it is meaningless to study the origin of the program independently of the metabolic network.

It is similarly erroneous to see the essence of life in proteins, or to look for the mystery of life in the spontaneous genesis of enzyme proteins. Enzyme proteins obtained a specific role in connection with the essence and origin of life because of their particular properties as catalysts. However, it is known that RNA could also be an enzyme; furthermore, it is very probable that in the near future artificial enzymes will be prepared from non-biological macromolecules (i.e., not from proteins or nucleic acids).

Thus, at present, the concepts of protein and enzyme are distinguished. Obviously, it is possible to go even further and say that in this case the essence of life is hidden not in proteins but in macromolecules of enzyme properties, thus the genesis of these should be investigated. However, enzymes accelerate only chemical processes that are thermodynamically possible. It is true that the present enzymes of the living world accelerate reactions to such an extent that in their absence the given reaction step of the metabolic network can be regarded as not existing at all, as a reaction step taking place 10 or 100 million times slower practically does not exist from the standpoint of the metabolic network. The extent of this "present" acceleration makes control possible with an efficiency that conceals the efficiency of all other regulations. Therefore no system is able to compete today in the living world without this very effective possibility of control. At present, the metabolism of every living being operates with a complicated enzymatic regulation.

This efficiency of today's enzymes is obviously the result of a long process of evolution. If we could go back in this evolution early process, we would find less and less effective enzymes in metabolism as catalysts. On carrying out this backward extrapolation consequently, we could reach a state where the individual steps of metabolism are not accelerated catalytically at all. The question that such a system (the metabolism of which contains chemical reactions without any enzymatic acceleration) is capable of behaving as living or not has not been studied so far except by the author. René Buvet has proved that the reactions taking place enzymatically, in today's living systems, could proceed without enzymatic catalysis under abiotic conditions (Buvet, 1974). Studies connected with chemotons have proved, as we shall see later, that for the regulated and program-controlled self-reproductive operation the information contained in a sequence and its utilization in a catalytic way is not a precondition. The essence of life is hidden in the mode of organization of the system that is regulated catalytically by enzymes (which are determined by their monomer sequence. And the operation of which is controlled by the sequence of DNA of the hereditary substance through the synthesis of these catalysts). Thus the primary task is to search for the organizational mode of this system, then to clarify whether such systems could come into being under primordial conditions. Only after all that should we investigate whether the sequence-dependent catalytic regulation and control could be built onto these systems in the course of evolution.

Life Criteria

Alive, capable of living, dead, nonliving

In order to fulfill our tasks formulated previously we need a system of criteria by means of which it can be unambiguously decided whether a system can be regarded as alive or not. Before the selection and formulation of these criteria, however, the meaning of some concepts used inconsequently should be unequivocally clarified.

As the counterpart of life it is usual to think of death. However, this is not quite true. First, in dehydrated living beings that can survive dehydration (e.g., spores, dried-up bacteria or protozoa), no life processes take place at all; thus they cannot be considered living, but neither are they dead, as they become alive on hydration. Thus the counterpart of the living state is the state without life, which may be the dead state, but may also be a state "capable of living," yet, without life. In addition to living beings in a dried-up state (but not dead), we have frozen living organisms (capable of living), living beings in the state of clinical death, seeds (in their dormant states), spores, etc., all of which can also be regarded as being in this state. For the state "capable of living" but not alive

Slemmer suggested the name β-state (Slemmer, 1981). In what follows, we also use this name.

Second, death may occur not only in the living state, but also in this β-state. In fact, this is, under certain conditions, even necessarily so. For example, the time course of the decrease in the number of "living" germs during dry sterilization is a well-known phenomenon, as well as the decrease in the percentage of germinations of stored seeds during long storage. Thus death means not only the cessation of the living state, but also that of the ß-state. At the same time, the β-state can cease to exist not only by death, but also by its counterpart: by "coming alive" as well.

Third, if we want to use the concept of death consequently, we should only consider dead what has died, i.e., what has been alive previously or at least has been in the β-state. (Everyday language, and sometimes even professional terminologies use the expression "dead" arbitrarily for systems and materials of the nonliving world e.g., for those of inorganic nature, though they have not died because they have never been alive).

Thus a biological individual may be in one of the three states:

1. It may be alive (let us call it the α-state), in which the particular characteristic processes take place;

2. In the state capable of living, but not alive (β-state) in which these specific processes do not take place, but they can start if the necessary conditions are provided;

3. In the dead state (let us call it the γ-state), in which certain processes may occur an one or another of them may be identical with those in the living state, but in which the order of processes characteristic for the α-state is not present and from which the system cannot be transferred into either the α- or the β-state.

The "capable of living" state may be either the α- or the β-state. Consequently, death terminates not the living state, but the capable of living state. The dead state is the counterpart of the capable of living state. Thus systems that have never been capable of living (e.g., inorganic systems) cannot be dead.

The nonliving state may also be of three kinds: the β-state, the γ-state, or the originally nonliving state.

The counterpart of living is nonliving; that of capable of living is dead. Death means not the cessation of life, but that of the capability of living.

The levels of life

According to the previous discussion we can establish that biology is not the science of living systems but that of the capable of living systems. In other words we may also say that biology deals with mortal systems that may be in a living or nonliving (but not dead) state. Based on this, we can define our task even more precisely: we should look for fluid (chemical) automata that reveal properties characteristic of living in their operating state, plus, those of capable of living but nonliving in their dormant state. Thus the borderline is mortality: the fluid chemical automata we are looking for should have an internal organization, the cessation of which has the ceasing of the capability of life as a consequence.

However, neither is the phenomenon of death unambiguous in the living world. The concept of death developed as the cessation of the life of vertebral animals. The ceasing of life for both man and vertebrate is instantaneous (i.e., with respect to the lifespan of man or animal an event taking place in a negligible time). This is transferred into phraseology concerning this event: we speak about the moment of death.

At the level of vertebrate the concepts of life and death are unequivocally connected to the entity: no vertebrate can be sectioned into two parts so that both halves would remain alive. The situation is similar for insects, spiders, or crabs, and in general for animals having a brain. With these living beings, the cessation of capability of life, i.e., the setting in of death may be detected unambiguously by the ceasing of the operation of the central nervous system.

However, there have already been problems connected with the death of lower animals in the past century. For example, annelids may be cut in pieces so that a full entity develops from every piece – usually an undamaged segment is enough for the development of a whole animal. Hydras and poriferas may be sectioned almost without any limits – from one single cell of poriferas a whole animal develops.

In the 1930s century some contradictions arose concerning the death of vertebrate as well. The tissue culture technique developed by Carrel has proved that the cells and tissues of animals may be kept alive after their death and may even be propagated. As the setting in of death for vertebrate can be detected unambiguously, and death means not only the cessation of life but also that of capability of life, the phenomenon of "surviving" tissues and cells can only be interpreted in that the animals have life at two levels; on one hand they live at the level of their cells, i.e., that each of their cells is a separate living and mortal unit, and, on the other hand, they form a system built of these units. The multicellular vertebrates built of billions of cells also possess properties at this higher level of organization, which may also be called life and is characterized by mortality.

Mortality – and thus also its counterpart, capability of life – is the characteristic of a system unit. Such a unit is the vertebrate, but such units are also the cells constituting the vertebrate. If we consider the life of the cell to be an organized system of chemical reactions in the chemical state space (as a first approximation), and this mode of organization is regarded as responsible for the properties characteristic of life (the life of the cell), the ceasing of which results in the death of the cell, then the functioning of the multicellular animal (i.e., its life) should be regarded as the organizational mode of the operation of a community of cells. This organizational mode is then responsible for the properties characteristic for the life of the animal as a multicellular unit, and its cessation results in the death of the multicellular animal even if the cells as units are living further and there is no change in their organization.

Thus we should look for a mode of organization that can be found at the biochemical level of the cell, but also separately at a higher level of organization, at the level of the multicellular animal effected by the coordination, regulation, and control of operation of cells.

The situation is not at all so unambiguous for other multicellular animals such as, for example, animals with a ganglion chain or with a diffuse nervous system, or for multicellular plants. In their case we look in vain for the phenomenon of death at two levels; it does not exist. A porifera or a plant cannot be killed without killing their cells at the same time; they die if and only if their cells die simultaneously.

If the phenomenon of death at two levels does not exist for animals not possessing a brain, the possibility of double life does not exist either: these multicellular organisms are systems organized from living units (the cells) the supercellular organization of which does not possess the features characteristic for life. They are also, of course, living systems, but only because their cells live. They die if their cells die, and these multicellular organisms die because their cells die. However, animals possessing a brain can die without the death of their cells. Moreover, when these animals die, their cells die because the multicellular organism died.

The difference between these two types of multicellular organisms is that those being alive at two levels operate with a central control at two levels; that is the operation of every one of their cells is controlled centrally by the cells' own genetic material, and the system consisting of these cells is centrally controlled by the brain. In the cells of living beings alive only at one level, this central control is carried out by DNA, but the operation of the system consisting of these cells is not controlled by a central organ. Thus, according to the previous discussion, in order to consider a system living at the given level of organization, an indispensable requirement is the presence of a central controlling subsystem. In the present form of the cell, this control function is carried out by DNA through its nucleotide sequence; thus for the life of today's cells DNA is indispensable. In the organism of higher animals this control function is carried out by the brain; thus brain function is an indispensable part of life at the second level in these animals. This is why it is justified to consider the existence of brain function the criterion of the living state.

Life criteria

The totality of life processes, i.e., the operation of living systems, results also in phenomena suitable for the characterization of life itself at both levels of organization. There are phenomena to be found among them, in every individual *in every phase of its life,* for with the continuous and joint presence is necessary, or the system does not live. Thus their continuous and joint presence is an indispensable criterion of the living state; therefore they will be called *absolute (real) life criteria.* However, there is also a group of life criteria the presence of which is not an indispensable criterion of the momentary living state of individual organisms, but they are indispensable for the evolution and survival of the living world. They are also regarded as life criteria but unlike the former ones, they are called *potential life criteria.*

Though very arbitrarily, biology has already used criteria for differentiating between living and nonliving. These "classical" life criteria used to be called *life phenomena.* It can be shown that these so-called life phenomena are not suitable for telling living and nonliving apart. Here the criticism of classical life phenomena will not be dealt with; the reader therefore is referred to the earlier works of the author (Gánti, 1971, 1977, 1978b, 1979a, b).

Based on the above, the following will be regarded as life criteria (Gánti, 1971, 1978b, 1979a).

Absolute (real) life criteria

1. **The living system should be an inherent entity**

   A system is considered an entity if its properties are not simple sums of the subsystems properties, and if the whole entity cannot be divided into its subsystems so that the subsystems carry the properties of the whole system.

   The system forming an entity is not a simple union of its elements, but a new quality that carries new properties with respect to those of its parts. These new properties are determined by the interaction of the elements of the system according to its organization. Only the system as a whole possesses the totality of these properties.

   Biological systems form entities inherently, i.e., as a consequence of their inner essence. The statement that these systems form entities inherently does not contradict the fact that the system may have additional parts, i.e., entities do not have to be minimum systems.

   As both the design and function of biological entities are genetically determined, genetic code carries information for the given living entity. For example, the DNA of a multicellular animal carries the information pertaining to itself as a macromolecule (see Chapter II, Section 5), and also that referring to the cell as a living entity; moreover, it contains information that is not indispensable for the life of the cell or for the DNA, but is necessary for the life functions of the multicellular animal as a secondary living system.

2. **Living systems should have metabolism**

   By metabolism we mean the process in which material and energy from the environment enter the system actively or passively, the system converts them chemically to its own matters with the formation of by-products, the chemical reactions lead to the controlled internal growth of internal matter and/or to the supplying of energy to the system, and finally metabolic waste materials leave the system actively or passively.

   The criterion of metabolism is essentially equivalent to the statement that living systems are fluid (chemical) automata.

3. **Living systems should be inherently stable**

Inherent stability is identical neither with equilibrium nor with steady state (Gánti, 1979a, b). It means a specific organizational mode of the inner processes of the system that enables the system to remain essentially unchanged in spite of its continuous operation and, moreover, in spite of changes (within certain limits) in its surroundings. It means that though the system reacts continuously to the external effects through the dynamic change caused in the living system by the variation of the external parameters, it remains unchanged as a whole. It also means that in spite of the constantly occurring chemical transformations in it, the living system does not decompose; if necessary, it may even grow.

This inherent stability is more than homeostasis because homeostasis is its consequence. Inherent stability is an organizational property, which, as we shall see later, is a natural consequence of the network of elementary chemical and physical processes taking place in living systems. The dormant seed, the frozen tissue culture, the lyophilized microorganisms, or the dried worm or protozoa are neither in homeostasis nor in steady state, though all of them are provided with the criterion of inherent stability, which is manifested in the fact that under suitable conditions they become alive again.

Systems with inherent stability show homeostatic properties during operation; thus this criterion includes the criterion of homeostasis as well. A system with inherent stability does not show homeostasis in the β-state but it can be revived from this state and then it possesses the homeostatic property.

However, homeostasis, i.e., the maintenance of the internal milieu of living systems, may be realized only through perceiving changes in the environment and by active, compensating responses to them. Thus sensitivity (excitability) is the mode of realization of homeostasis, its mechanism. Thus inherent stability, as a life criterion, includes not only the criterion of homeostasis, but also that of sensitivity – excitability in its most general sense.

4. **Living systems should also have subsystems carrying useful information for the system as a whole.**

Every being carries information concerning its own design, origin, and function. However, there are also systems capable of carrying information independent of themselves. Examples of these are books, magnetic tapes, punched cards, records, etc.

In nature, only certain subsystems of the living systems, e.g., the genetic material, the brain, and the immune system, are capable of carrying "excess information" These contain information not only for themselves, but for the whole living system, and eventually also for the world outside the living system. The presence of information-carrying subsystems is the property of every living system without exception, and it is an indispensable criterion of the evolution of the living world.

As it has already been shown, coded information really becomes useful information only if another system capable of reading and utilizing it also exists. Thus it is also characteristic of living systems that they are capable of reading, and utilizing, and moreover, at the level of primary life during proliferation, also of copying, i.e., replicating this information. Not only is the storage of information characteristic of living beings but also the capability of informational operations.

5. **Processes in living systems should be regulated and controlled by a program.**

The condition of the existence of every dynamic, i.e., continuously operating, system is the regulated nature of its processes. Living systems as dynamic and fluid systems also possess this property. Regulation occurs mainly through chemical mechanisms in living systems. In fact, regulation itself should not be listed as a separate life criterion, as neither metabolism, nor homeostasis could be realized without the regulation of the processes in the system; thus these criteria automatically postulate the regulated nature of the system.

Regulation in itself is capable of maintaining only the existence and function of the system. Growth, proliferation, differentiation, and ontogenesis are irreversible processes that cannot be realized exclusively by regulation. For these processes program control is necessary in addition to regulation. Program control and the storage of the program take place in living systems by means of fluid (chemical) mechanisms.

Potential life criteria

6. **Living systems should be capable of growing and proliferating.**

Growth and proliferation are classified among classical life criteria, and their presence is general and indispensable in the living world. However, they are not criteria for the living state itself, as castrated or aging animals lose their capability of growing or proliferating without losing their life. Their presence is not a criterion for individual life, but it is for the living world.

Growth and proliferation are contracted into a single criterion, as they became separated only at a certain level of phylogenesis; the two processes are inseparable for prokaryotes or for a part of eukaryotic cells. This was examined and revealed by Hartmann, who operated daily for 130 days on an amoeba, which was normally dividing every second day, so that he removed one third of the cytoplasm each day while still leaving the nucleus untouched. The amoeba did not die, but neither did it divide during those 130 days because it never reached the growing stage needed for division (cited by Huzella, 1933).

The proliferation of multicellular organisms is only indirectly connected to their growth, but the latter is mainly a consequence of the proliferation of their cells.

7. **Living systems should possess the ability for hereditary change and evolution**.

   By inheritance it is meant that living systems are capable of producing individuals identical with, or similar to, themselves or some germs ensuring the production of such individuals. Consequently, the concept of proliferation also includes inheritance.

   The living world could not have evolved if inheritance were strict, i.e., if the properties of descendants were identical with those of the parents or with the combination of their properties in every respect. Therefore the ability for hereditary change, i.e., the fact that in the descendants properties may appear that were not present in the infinite chain of the series of predecessors, should be listed as a separate criterion.

   The ability for hereditary change is a necessary but probably not sufficient condition for the evolution of the living world. For evolution, the possibility of the appearance of nonequivalent variants is also necessary. Criterion 7 expresses this possibility in living systems.

In certain works of the author (Gánti, 1971, 1977) mortality is also mentioned as a potential life criterion. Namely, in addition to the already discussed confrontation between mortal and living, it is obvious that for the survival of the living world the mortality of individuals is an indispensable criterion, as only then may the matter of living beings return into the mass cycle of the biosphere.

This is also a necessary condition for the stability and evolution of the living world through the production of newer and yet newer individuals.

The criteria listed include also the classical life phenomena (except for movement) but in a stricter and more well-defined form. Besides, there are further criteria among them the life-criterion character of which could be formulated only in light of the results of molecular biology (e.g., the regulated–controlled nature of informational operations and processes). Thus the life criteria listed here impose much stricter limitations on a system to be considered living than the classical life phenomena.

As the presence of potential life criteria is not a precondition of life at the individual level, every system satisfying the absolute (real) life criteria should be considered living, independently of the actual design of the system or the chemical nature of the materials that constitute it. However, for the formation of a living world only those systems are satisfactory those fulfill both the absolute and the potential life criteria.

Such a definition of life makes the discussion of the basic laws of life possible in a fully generalized manner, independently of the actual material quality or modes of realization of the living systems.

As a test of this statement, let us see how biological systems at different organizational levels satisfy these criteria.

1. Individuality.

   Prokaryotes are surrounded by membranes (microplasms, thermoplasms) or by membranes and cell walls. Their uniform operation is ensured by the genetic material. Eukaryotes are surrounded by membranes (animal cells) or by membranes and walls. Their uniform operation is ensured by their nucleus (chromosomes). Animals possessing a brain are covered by epithelium and their uniform operation is governed by the brain. Plants do not generally satisfy this criterion at the level of the multicellular organism, but they do at the level of their cells.

2. Metabolism.

   At the prokaryotic level this is a purely biochemical organization. The intake of nutrients and excretion occur by physicochemical mechanism and active transport. In eukaryotes, a new organizational level appears above the biochemical one with the coordination of the functions of cell organoids (mitochondria, chloroplasts, lysosomes, cytoplasm, nucleus, etc.). New elements appear also in the intake of nutrients and excretion: mouth organs, secretory organs, endocytosis, and exocytosis. In addition to the previously mentioned levels, in multicellular animals yet another new level of organization appears: the ordering of the functions of different organs into a system (mouth organ, masticatory organ, stomach, intestinal tube, internal secretory organs, liver, kidney, respiratory organ, vascular tissues, etc.). Intake and excretion of food occur by the coordinated operation of organs that function at a higher level than the mechanisms operating in lower living beings. This criterion is also satisfied in plants, as the intake of nutrients, transport, photosynthesis, etc. are carried out by separate organs.

3. Inherent stability.

   This criterion is satisfied in prokaryotes by chemical mechanisms only. At the eukaryotic level, above the biochemical mechanisms, stability arises from the quantitative and functional coordination of cell organoids (e.g., acceleration or deceleration of the proliferation of mitochondria and chloroplasts, formation of vacuola, etc.). In multicellular beings at a level above all these, a system ensuring hormonal and neural stability also appears. The criterion is again also satisfied for plants.

4. Information-carrying subsystems.

   At the prokaryotic level, DNA carries the information. In addition, at the eukaryotic level chromosomes (as structural units) also do this. Again at a level above all these, in multicellular animals the brain (nervous system), or in specific cases the immune system, performs this. In plants it is satisfied only at the level of cells.

5. Internal control.

   At the prokaryotic level it is of a purely biochemical nature. At the eukaryotic level it is also biochemical but at a higher level of complexity. At the multicellular level, it is neural control. In plants it does not exist, or it is transferred from the program stored at the cellular level into the whole of the plant through the self-regulating processes of the supracellular organization.

Chemotons as Living Systems

Chemotons satisfy life criteria

If we accept the previous criteria as criteria of the living state, then every system satisfying absolute life criteria should be considered living, and every system satisfying in addition the potential life criteria must be considered as a living system capable of evolution into a living world. In Chapter IV it was stated that chemotons possess the basic properties of prokaryotic living systems and satisfy life criteria. Thus, in what follows, armed with the knowledge of life criteria and the properties of chemotons, this statement has to be proved. This earlier statement may now be completed in the sense that chemotons satisfy both the absolute and potential life criteria. Thus they should be considered not only living systems, but also systems from which, by evolution, a complex living world can develop.

a) **Chemotons form inherent units, as their properties cannot be obtained by the simple addition of those of their subsystems.** The simplest variants of chemotons are composed from only three subsystems: from the autocatalytical chemical cycle (or reaction network), the template polymerization system, and the growing membrane. From the mode of coupling of these subsystems new properties appear that were not possessed by the separate subsystems. Program control appears as a result of the stoichiometric coupling of the autocatalytic network and template polymerization, whereas the ability for proliferation appears as that of the autocatalytic network and membrane growth. The stoichiometric coupling of the three systems results in the ability for hereditary change appearing as a new property as is shown later in detail. Thus these properties originate from the organization of the chemoton, i.e., they are inherent properties originating from the inner essence of chemotons.

Nevertheless, the previously mentioned properties are only the verbal formulation of the very complex behavior of a system. In fact, only computer simulation can show that the system behaves as an integral whole and that it responds to external stimuli as such (see for example Figs. 4.7.–4.9). This uniform behavior is best illustrated, however, by the stability tests already discussed in connection with the stability criterion. Thus chemotons satisfy the first absolute life criterion.

b) **Chemotons have metabolism as they transform nutrients entering from their surroundings into internal material in a chemical way, resulting in a regulated and controlled growth of the total of their internal materials.** As has been already established in the discussion of life criteria that the criterion of metabolism is synonymous with that of living systems being fluid (chemical) automata, and as chemotons have been designed as the proliferating variants of fluid (chemical) automata, this criterion is automatically satisfied for chemotons.

c) **The stability of chemotons originates from the mode of their internal organization, i.e., it is inherent.** None of the subsystems of chemotons are stable in themselves because they are monotonously growing as a result of their autocatalytic nature, and no monotonously growing system may be stable. However, the stoichiometric couplings between these systems may interrupt this monotony; moreover, they may temporarily transform increase into decrease, as has been shown for oscillatory systems in Chapter III. The monotonous nature of the growth of subsystems may be destroyed and transformed into a complex periodic behavior by stoichiometric couplings also in chemotons, as is shown in Fig. 4.7 (obtained by simulation). However, the whole of the chemoton remains a growing system, i.e., when it operates the sum of its internal materials increases continuously even though the amounts of some individual components may decrease perhaps drastically (e.g., components R or T^*). According to this, the chemoton as a whole could not be a stable system. The problem is solved by the division of the chemoton, namely, when the chemoton, in its adapted state where the amount of its components is doubled with respect to that of its starting state, divides into two chemotons identical with each other and with the original chemoton (Gánti and Gáspár, 1978). Thus in every individual chemoton the original state is restored, as is seen on the different curves in Fig. 4.7.

   If the changes occurring between two divisions are plotted as a concentration versus concentration curve (i.e., the concentration changes of the various components are represented as a function of the concentrations of each other in pairs), instead of the periodic curves of Fig. 4.7 we obtain limit cycles, which are traversed by the adapted chemotons generation after generation, and which are approximated by chemotons during adaptation. Homeostasis does not mean the maintenance of the values of individual parameters but rather that these parameters should always wander on the limit cycles even though the course of these limit cycles changes with the variation of the external parameters. The variation of internal concentrations or amounts does not change the limit cycle but only places the system outside it; however, the system will return into it automatically during the course of operation. An exception is the variation in the amount or composition of the template polymer, and as we see later, hereditary change is realized through this phenomenon.

These stability properties of the chemoton as a system originates from the internal organizational mode, i.e., from the inner essence of the chemoton, as its subsystems are in themselves not stable. Thus chemotons are inherently stable systems revealing homeostasis during operation.

There is also the question as to whether or not there exists a nonfunctioning, but capable of operation, state of chemotons (or in other words, can they enter the β-state?). The answer, of course, is that they can. Such a state is realized if the thermodynamic driving forces operating chemotons cease to exist. This may occur in two ways: (1) in the case when all the nutrients are consumed, or (2) if the concentration of waste materials increases to such an extent that the reactions of the cycle reach the equilibrium state. This latter case can occur if for cycle Eq. [4.58] the following condition is fulfilled:

$$v^{*} > \frac{Kx}{yt'(e)\cdot a_1(e)} \qquad [6.5]$$

where $K = \prod_{i=1}^{5}\left(k_i / k_i^{*}\right)$ and $t'(e)$ and $a_1(e)$ are the equilibrium concentrations of T and A_1 (Békés, unpublished results). This is the immobile state originating from the kinetic equilibrium state and from which the system is displaced by even the smallest variation in x or y.

The situation is different when the nutrients are depleted. It can be seen from Fig. 4.6 that the operation of the autocatalytic cycle cannot be reversed because of the irreversibility of the transformation $T' \rightarrow T^{*}$, thus the value of $\sum_{i=1}^{5}\alpha_i A_i$ cannot decrease. As there is no stoichiometric tapping both in the cycle of pV_n and in membrane growth, all of the subsystems constituting the system remain capable of operation, and in the presence of nutrients the systems start to operate again. Thus chemotons with a stoichiometric coupling as depicted in Fig. 4.6 have stable β-states. This stability in the β-state also originates from the inner mode of organization of chemotons, i.e. is also of an inherent nature.

d) **Chemotons are provided with a subsystem carrying information for the whole of the system (as well), and this is the template polymerization system.** In order to prove this, we have to go back to the system described in Chapter III, in which the autocatalytic cycle is shown to be coupled to the template polymerization process by an AND-coupling. It is shown there that the length of the template (and for co-polymeric templates also the composition of the template) also carries information for the autocatalytic cycle in the coupled system, as it controls the operation of the system in practically the same way as a program. As in the chemoton shown in Fig. 4.6, the self-reproductive cycle is coupled to the template reaction with the same coupling, and this relationship also holds inside the chemoton. However, in chemotons the component liberated in the template reaction as a condensation by-product is an indispensable reaction component for membrane formation; its amount determines unambiguously the amount of membrane-building material formed and through this the possibility of an increase in membrane surface during one generation time. (The amount of the other reaction component is not so restricted because of the possible shift in reaction equilibrium.) Therefore information concerning the extent of membrane growth in chemotons is carried by the length of the template (or the total length of the templates).

Thus the template subsystem in chemotons carries information for its own replication, for the operation of the cycle, and for membrane growth as well. This information-carrying role is discussed in more detail when we examine the genetic properties of chemotons. It should also be noted here that in more advanced variants of chemotons the information-carrying role of the template is realized not only through the length of the polymer, but also through its composition, and finally, also through its sequence.

e) **Processes in chemotons are regulated and program controlled.** This statement has already been proven in Chapter IV. by showing that, in chemotons, processes are regulated by complicated interactions between amounts, concentrations, and volume changes induced by the increase of the surface are, and that they are controlled in their qualitative behavior as well as in their time course by the template process.

On the basis of all the preceding points it can be established that chemotons satisfy absolute life criteria, and if we accept these criteria for telling living and nonliving things apart, systems with chemoton organization should be considered living systems. As it has already been derived that these systems possess the ability for growth and proliferation, and as it is shown later, in discussing the genetic properties of chemotons, that they are also provided with the ability for hereditary change and evolution, chemotons may be regarded as living systems capable of serving as the basis for the development of the living world.

Chemotons are minimum systems of life

If the chemoton model is considered a model of living systems, the problem should be examined as to whether it is possible to create a simpler model satisfying life criteria or – conversely – chemotons are the simplest systems that may be regarded as living.

In the last time three models have been published, the internal organizations of which can be described by exact methods and that are considered living by their authors: the bioid model by Decker, the hypercycle by Eigen, and the chemoton model. In addition, several experimental models are known (the microspheres of Fox, the Jeewanu by Bahadur, etc.) and also some concepts are described qualitatively (e.g., the concept suggested by Hans Kuhn). However, because they lack an exact quantitative description, they will not be dealt with here but rather later in following chapters.

The simplest concept also treatable quantitatively is that by Peter Decker (Decker, 1974; Decker and Speidel, 1972). This is essentially a simple autocatalytic cycle, or two different autocatalytic cycles coupled by the branches of two branched reaction chains (Fig. 6.1). According to Decker, the formose cycle behaves as a bioid. However, it is obvious that bioids are not capable of satisfying even the absolute life criteria, for they are not inherent units, they do not possess information-carrying subsystems, they are not program-controlled, and they are not inherently stable. At the same time, they seem to be suitable for acting as metabolic subsystems inside living systems.

The hypercycle model constructed by Manfred Eigen (Eigen, 1971; Eigen and Schuster, 1977) is significantly more complex: it is an autocatalytic supercycle, the subsystems of which are autocatalytic template cycles themselves (Fig. 6.2). This construction obviously satisfies the criterion for the information subsystem through the I_i-s, as well as that of program control. However, it does not satisfy the criterion for inherent entity because of the lack of compartmentalization or that of inherent stability from the absolute life criteria.

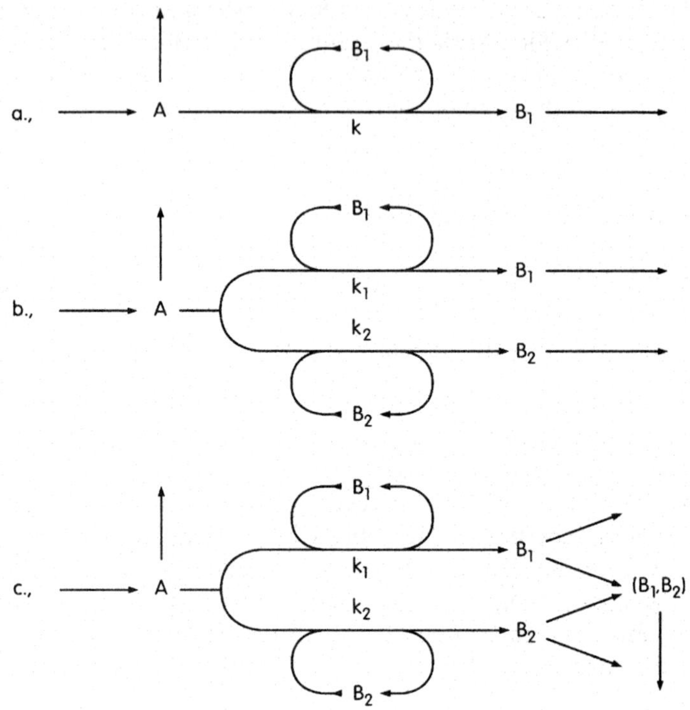

Fig. 6.1. Various "bioids" shown according to the original description of Decker (1974). a) Simple bioid; b) competitive bioid; c) aggressive bioid.

Eigen referring to Oparin, regards metabolism, self-reproduction, and the ability for mutation as life criteria, and he finds that the hypercycle satisfies these criteria (Eigen, 1971). However, as we have shown already, self-reproduction cannot be identified with proliferation including also spatial separation; thus the hypercycle does not satisfy one of the potential life criteria either.

Thus the hypercycle cannot be considered a living system according to the previously discussed system of life criteria. It corresponds rather in its function to the information-carrying subsystem of the chemoton, and yet it is more complex than this.

Finally, the question arises as to whether it is possible to leave out some of the subsystems from the chemoton in such a way that the remainder still satisfies the life criteria. If the membrane system is removed, the program-controlled fluid

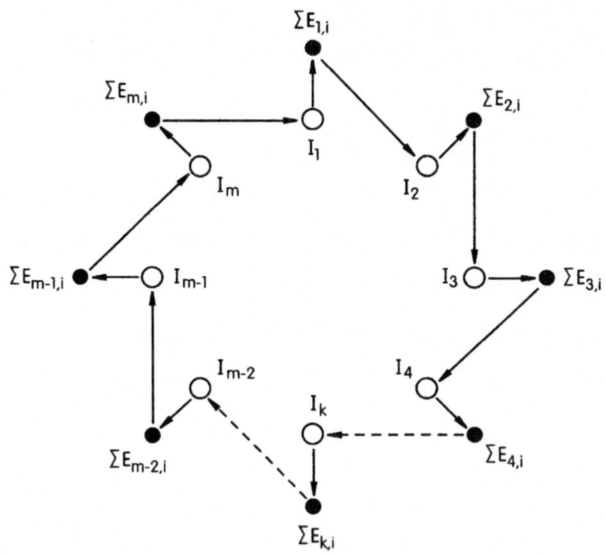

Fig. 6.2. "A self-instructive catalytic hypercycle" according to the first work by Eigen about hypercycles (Eigen, 1971). I_i-s are information carriers (e.g., RNA strands). E_i-s are catalysts coded by the I_{i+1}. Each E_i branch is capable of performing different functions (e.g., polymerization, translation, control); one of them is to aid specifically the formation of I_{i+1} because of its specific ability for recognition.

automata (Chapter III), which is not an inherent unit, does not satisfy the criterion of inherent stability and, in addition, is not capable of proliferation (though it is self-reproductive). If the information-carrying subsystem is left out, program control and capability of evolution cease, and we arrive at the simple proliferating microsphere (Chapter IV). If, in turn, the autocatalytic cycle is left out, then it should be ensured, for the operation of the two other subsystems, that the membrane be penetrable for their raw materials (T^*, V^*) and that in the extrachemotonial space they be present in a sufficiently high concentration. However, in this case, because of the lack of osmotic regulation, the ability to divide ceases to exist, and therefore neither the criteria for inherent stability nor ability for proliferation is satisfied.

Thus, based on our current knowledge, the chemoton model should be considered the minimum model of living systems. It is not, of course, proven that this is the only possible model, i.e., that it is not possible to construct another model capable of satisfying life criteria. However, it is shown later that the operation of every living system on Earth is based on chemoton coupling, according to our present knowledge.

VII. Chemical Evolution

Setting of Objectives

The aim of this and the following chapters is not to provide a final and detailed answer to the question of the origin of life on Earth but to show that this problem may be solved in an exact manner by chemoton theory. If chemotons are the minimum systems of life – in our case, life on Earth – it may be righteously supposed that life has come into being through such systems. The fact that the properties of chemoton models can be studied both qualitatively and quantitatively in an exact way provides an opportunity to discuss the events of biogenesis qualitatively and quantitatively, along with checking experimentally the theoretical results thus obtained.

The use of the chemoton model in the study of biogenesis is made possible by the fact that the model has no restrictions concerning the concrete chemical nature of the components. Thus the network of the model may be filled up with compounds forming (or reactions taking place) under prebiotic conditions if they satisfy the requirements determined by the model, and on the other hand, if the model does not contain restraints for the complexity of its subsystems. They can be substituted by any real chemical system of arbitrary complexity if the latter are provided with the self-reproductive (self-productive) properties determined by the model and the necessary stoichiometric couplings can be established between them.

If we knew all the reactions taking place spontaneously under prebiotic conditions, we would need only to investigate in what manner and in how many ways metabolic networks may be designed, satisfying also the coupling conditions determined by the chemoton model. Thus, the chemoton model makes possible, in an exact, theoretical manner the design of alternate modes for the origin of life while being in possession of the basic data. It also makes possible the checking of the alternatives, theoretical correctness (by detailed stoichiometric and kinetic calculations) and viability (via directed and precontrived experiments).

The number of papers dealing with chemical events under prebiotic conditions, since the pioneering work of Stanley Miller (1953), is close to ten thousand. As a result, a number of prebiotic chemical processes have become known, as is shown later. Unfortunately, the direction of these studies has been partly inspired by experience originating from the present living world (e.g., the emphasis on proteins) and partly restricted by the available analytical methods. Thus the studies do not consider the different families of compounds with equal weighting. The majority of the results refer to the prebiotic origin of amino acids, followed by that of organic acids, nucleic acid bases, nucleotides, and polynucleotides. There are only very few data concerning the abiotic reactions of sugars and sugar phosphates, and hardly any concerning lipids and lipoids. Therefore the experimental data at our disposal do not make possible the design of a prebiotic chemoton metabolic network substantiated in all its details.

Nevertheless, the data do allow us to show that such a network may be designed and characterized by cycle stoichiometry describing it by an exact stoichiometric overall equation. These data are shown in this chapter. In the following chapter, a concrete metabolic network of a prebiotic chemoton is designed. The backbone of the network will then consist of prebiotic reactions already known, and this is completed with hypothetic chemical reactions in such a way that the final network obtained should be a full metabolic network of chemoton coupling. By "full metabolic network" it is meant that the system synthesizes all of its components by using several simple nutrient components so that, between two divisions, the amount of every internal component becomes doubled. After that it is shown how the stoichiometric calculation of this system can be made, and the appearance of enzymes and genes is derived.

The Date of the Genesis of Life

In order to outline the actual metabolic network of a prebiotic chemoton, the conditions have to be known under which they are assumed to originate, i.e., in our case, the prevailing conditions on primordial Earth should be known. However, conditions on Earth have changed many times since its genesis, partly as the result of the evolution of Earth as a planet, partly as the result of changes in the Sun as the energy source, and partly as the result of chemical, biological, geological, and meteorological processes taking place on the surface of Earth.

Thus, the conditions at the time of the genesis of life on Earth should be known, and to this, the date of the origin of life.

Unfortunately, this date is not known exactly. However, borderlines can be drawn before which life could not exist and after which life was surely existent on Earth. If, e.g., Oparin's concept is accepted, according to which life originated from primeval oceans (Oparin, 1957), life could not exist earlier than these oceans did. The situation is the same if Corliss's concept is true; that the so-called black smokers, the hot springs of several hundreds of degrees centigrade, were the source of life (Corliss, 1984). According to Sidney Fox's concept, life originates from rain containing organic materials and falling onto hot rocks (Fox, 1965, 1973c). In this case, the origin of life preceded, or at least may have preceded, the formation of primeval oceans. Finally, if Woese's concept is considered correct, life was formed in the rain-drops floating above the still hot, primordial Earth, and therefore the origin of life could precede even the formation of a solid crust (Woese, 1979).

Earth is generally supposed to have been formed 4.6 billion years ago. The oldest pieces of rock found are particles of zircon ($ZrSiO_4$) and they have been assessed to be 4.1–4.2 billion years old (Williams and Myers, 1983). They have been found in secondary rock conglomerates formed from the detrita of earlier existing rocks about 3.5 billion years ago. Thus 4.1–4.2 billion years ago there were surely solid rocks existent on Earth. The oldest rocks have been found on the west coast of Greenland (Ishua); they are of volcanic origin and were formed under complex geological conditions with traces of a strong hydrothermal activity. They are 3.8 billion years old. In quartzites, some microfossil-like formations were found (Pflug and Jaeschke-Boyer, 1978), the biological origins of which are, however, strongly debated (Bridgwater et al., 1981; Schöpf, 1976). In the same rocks, 3.8-billion-year-old amino acids and hydrocarbons have been detected (Nagy et al., 1981). Thus the beginning of hydrothermal activity and the geological accumulation of organic material goes further back than that time.

The oldest stromatolites are about 3.5 billion years old (Lowe, 1980; Walter, Buick, and Dunlop, 1980). They are mats of limestone, formed by the activity of photosynthesizing microorganisms from the shallow seawater, and they often contain microfossils.

The oldest microfossils of a definite biological origin are from South Africa, Swaziland (Fig Tree formation), and from West Australia (North Pole); they are 3.1–3.5 billion years old, and their shapes are very diversified (Barghoorn and Schöpf, 1966; Brooks and Shaw, 1973; Groves, Dunlop, and Buick, 1981; Knoll and Barghoorn, 1977; Schöpf and Barghoorn, 1967;). J. William Schöpf, one of the best experts of early fossils, estimated in his lecture at the ISSOL Conference in 1983 that the origin of life can be dated back to earlier than 3.5 billion years and that anaerobic photoautotrophy is probably even older, whereas aerobic photosynthesis began to form about 2.9 billion years ago (Nagy and Zumberge, 1976; Schöpf, 1983).

On the other hand, in 1984 at the Cospar Congress in Graz, Schidlowski reported that, based on the measured distribution of C-isotopes, the conclusion has to be drawn that 3 billion years ago photosynthesizing systems operating with ribulose-1,5-diphosphate carboxylase should have existed. It is known that at present the $^{12}C/^{13}C$ isotope ratio differs significantly in organic and inorganic (carbonate) carbon compounds. It has been established that enrichment in ^{13}C is the consequence of the operation of the RuBP carboxylase enzyme participating in photosynthesis. The shift in the isotopic ratio can be followed back to at least 3.5 (and maybe to 3.8) billion years in the fossiled organic material (Chang, Lennon, and Vollner, 1977; Schidlowski, 1984).

If the enrichment in ^{13}C is really the consequence of the operation of the RuBP carboxylase enzyme, then 3.5 billion years ago not only living beings, but also photosynthesizing organisms using RuBP enzyme had to exist. This seems to be supported by the early appearance of stromatolites. However, photosynthesis using RuBP carboxylase enzyme means a very high grade of evolution, which could only come into being as a result of a long evolution process. We are therefore probably not very much mistaken if we suppose that life on Earth appeared 4 billion years ago or perhaps even earlier.

The Genesis of Earth

The number of measured data providing direct information on the surface conditions of primitive Earth is negligibly small; thus the situation on primeval Earth may only be derived indirectly from the theories concerning the origin of the solar system. These theories are based mainly on physical and astronomical considerations and they agree that our solar system has been formed from an interstellar dust and gas cloud (referred to here after as "interstellar cloud" or "cloud").

According to our present knowledge, there appear to be two basic types of interstellar clouds: high-temperature (30–100 K), diffused atomic clouds, and more condenser, cooler (10–20 K) molecular clouds. Diffuse clouds contain 10–10^3 atoms/cm^3; they consist of atomic hydrogen (in about 90%) and of helium

(10%). There are other components besides these (about 1%), e.g., microscopic dust particles (Blitz, 1982).

The main part of the denser, cooler (10^3–10^6 particles/cm^3) clouds is also hydrogen but 99% of it is in the molecular form. Other elements (C, O, N, S, Si, etc.), except noble gases, are present in molecular state as well. More than 60 different molecules, ions, and radicals have been identified in these clouds since 1968 (Blitz, 1982; Mann and Williams, 1980; Matthews, 1984; Turner, 1980; Zuckermann, 1977). Interstellar molecules identified up to 1984 are shown in Table 7.1. Stars, planets, and whole solar systems are formed from these more condensed clouds; therefore they are dealt with in greater detail here.

Name	Formula	Number of atoms
Cyanogen radical	C≡N	
Methylidine	C—H	
Methylidine ion	C—H^{+}	
Hydrogen	H—H	
Hydroxy radical	O—H	
Carbon monoxide	C=O	
Carbon monosulfide	C=S	
Silicon monoxide	Si=O	2
Sulfur monoxyde	S=O	
Silicon monosulfide	Si=S	
Nitrogen monosulfide	N=S	
Diatomic carbon	C=C	
Nitric oxide	N=O	
Carbon monoxide ion	C=O^{+}	
Water	H—O—H	
Hydrogen cyanide	H—C≡N	
Formyl ion	H—C=O^{+}	
Hydrogene isocyanide	H—N=C	
Carbonyl sulfide	O=C=S	
Hydrogen sulfide	H—S—H	3
Sulfur dioxode	O=S=O	
Ethynyl radical	H—C≡C	
Imidyl ion	N=N—H^{+}	
Formyl radical	H—C=O	
Thioformyl ion	H—C=S^{+}	
Nitroxyle	H—N=O	

Ammonia	H H—N< H	
Formaldehyde	H > C=O H	
Thioformaldehyde	H >C=S H	4
Hydrocyanic acid	H—N=C=O	
Cyanoethinyl radical	N≡C—C≡C	
Isothiocyanic acide	H—N=C=S	
Acetylene	H—C≡C—H	
Protonated carbon dioxide	H—O—C=O$^+$	
Tricarbon monoxide	$^-$C≡C—C≡C$^+$	
Cyanoacetylene	H—C≡C—C≡N	
Formic acid	/ O—H H—C=O	
Methylenimine	H >C=N—H H	
Cyanamide	H >N—C≡N H	
Ketene	H >C=C=O H	5
Butadynyl radical	H—C≡C—C≡C	
Methane	H | H—C—H | H	

Methylcyanide	H H—>C—C≡N H	
Formamide	H >N—C=O H \\ H	6
Methanol	H H—>C—O—H H	
Methyl mercaptan	H H—>C—S—H H	
Acetaldehyde	H H—>C—C=O H | H	
Methylamine	H H H—>C—N< H H	
Vinyl cyanide	H H | >C=C—C≡N H	7
Methyl acetylene	H H—>C—C≡C—H H	
Cyanodiacetylene	H—C≡C—C≡C—C≡N	
Methylformate	H H—C—O—C<—H ‖ H O	
Cyanomethyl acetylene	H H—>C—C≡C—C≡N H	8

Ethanol	H H H—>C—C<—H H O—H	
Dimethyl ether	H H H—>C—O—C<—H H H	9
Ethyl cyanide	H H | H—>C—C—C≡N H | H	
Cyanotriacetylene	H—C≡C—C≡C—C≡C—C≡N	
Cyanotetraacetylene	H—C≡C—C≡C—C≡C—C≡C— C≡N	11
Cyanopentaacetylene	H—C≡C—C≡C—C≡C—C≡C—C≡C—C≡N	13

Table 7.I. Interstellar molecules (identified up to 1984).

Upon studying the data in Table 7.1 it is apparent that the compounds shown are very aggressively reactive so the majority of them occur on Earth only as reaction intermediates of extremely short lifetimes. The reason for this is to be found in the unusually low temperature rather than the low density of the gas. As we shall see later, at this temperature even the most aggressive radicals freeze. From this fact it can be concluded that if the conditions causing this freezing cease to exist, violent chemical reactions resulting in very complex molecules would have to take place.

The other striking feature in Table 7.1 is the great number and diversity of components containing C, O, N, and S. This means that as a result of the previously discussed chemical processes the formation of a great number of complex organic compounds may be expected. Thus if solar systems are formed from the material of such clouds, the appearance of organic compounds on the planets of these systems is inevitable.

Data presented in Table 7.1 are the results of radio-astronomic and spectroscopic measurements on faraway clouds. Comets, we assume, gather their material from the cosmic matter of clouds surrounding the solar system. Their study can provide direct proof concerning the composition of these clouds. Unfortunately, the project for the examination of Halley's Comet has only just been started; therefore there are no direct measurements available as yet. However, the presence of about 40 components could already be detected by spectroscopic measurements more or less accurately (Table 7.2) and their relative frequency could also be determined (Table 7.3; Donn, 1982). These results are in accordance with those of radio-astronomic measurements.

The dust particles of the majority of clouds originate from M giants and super-giants; they are condensed in the atmosphere of these giants and propelled into space by the pressure of radiation. In space, so-called "dirty ice" (a mixture of methane, ammonia, and water) is condensed onto these silicate cores. At this temperature water is condensed in an amorphous state, transforming into crystalline ice only at about 100 K.

Atoms	Molecule	Ion
H	CH	Ca^+
C	NH	C^+
O	OH	OH^+
S	C_2	CN^+
Na	CH	CH^+
K	CO	CO^+
Ca	CS	N_2^+
V	NH_2	CO_2^+
Mn	C_3	H_2O^+
Fe	H_2O^*	
Co	HCN^*	
Ni	CH_3CN^*	
Cu	Silica	
Cr		

Table 7.2. Components of comets detected by spectroscopic measurements (*not confirmed).

Element	Relative frequency
H	1.5
C	0.2
N	01.
O	1.0
S	0.003

Table 7.3. Relative frequency of elements on the comets.

Concerning the type and development of these particles, Greenberg (1984) elaborated the following hypothesis based partly on spectroscopic data and partly on laboratory experiments. Because of radiation, the molecules in the primary mantle around the silicate core decompose, resulting in radicals that partly recombine into molecules of new types and are partly retained as frozen radicals, accumulating as such. Among the primary products of this process a prominent role is played by formyl radicals, formaldehyde, and glyoxal. Formaldehyde can polymerize even at about 20 K (Goldanski, 1977). With the increasing number of grains a diffuse cloud is formed that results simultaneously in an increase of the temperature in the particles. This heat formation triggers the recombination of radicals that may lead to such violent reactions that the mantle itself explodes.

No explosion occurs with a slow heating process, but in this case a series of complex processes takes place. As an end result, the material of the mantle transforms into so-called "yellow stuff," this being a stable mixture containing a great number of complex organic molecules (which have not yet been investigated in detail). As the cloud cools down and is condensed onto this yellow mantle, new dirty ice is deposited with a manifold of radicals and organic substances (Fig. 7.1). Such particles (consisting of a silicate grain, an inner mantle of yellow stuff, and an outer coat of dirty ice) are the starting phases for the formation of celestial bodies and solar systems.

The current theories concerning the genesis of the solar system agree insofar that solar systems are formed from interstellar clouds, but they differ as to the forces governing this process. One group of these theories is based on Laplace's theory, according to which the agglomeration is governed by gravitational and mechanical forces. In hypotheses based on the theoretical work of Alfvén, great importance is ascribed in addition to gravitational forces, to electromagnetic forces, and to electromagnetic processes taking place in the clouds. From our point of view, the great difference between the two groups of theories is that in the first case the formation of the solar system starts from a cloud of homogeneous distribution, whereas in the second case the formation starts from a chemically differentiated, inhomogeneous cloud. As the density distribution of the planets in the solar system (Table 7.4) and recent discoveries (e.g., critical speed, magnetic plasma interaction, the rings around Uranus and Jupiter) seem to support rather the theory of Alfvén, this is briefly described here (Alfvén, 1980a, b; Alfvén and Arrhenius, 1976; Alfvén and Fälthammar, 1963).

In cosmic clouds, regions with active electrical current and passive regions with negligible current alternate each others. Currents influence strongly the movement of dust particles and consequently a chemical differentiation starts in the cloud. As a result, the cloud becomes inhomogeneous, having a filamentary structure. This inhomogeneity may condense the cloud to such an extent that it reaches the threshold for gravitational collapse. The inhomogeneous current density may thus result in the collapse of the cloud in certain regions and in the formation of star-seedlings (the so-called stellesimals), and planet-seedlings (the

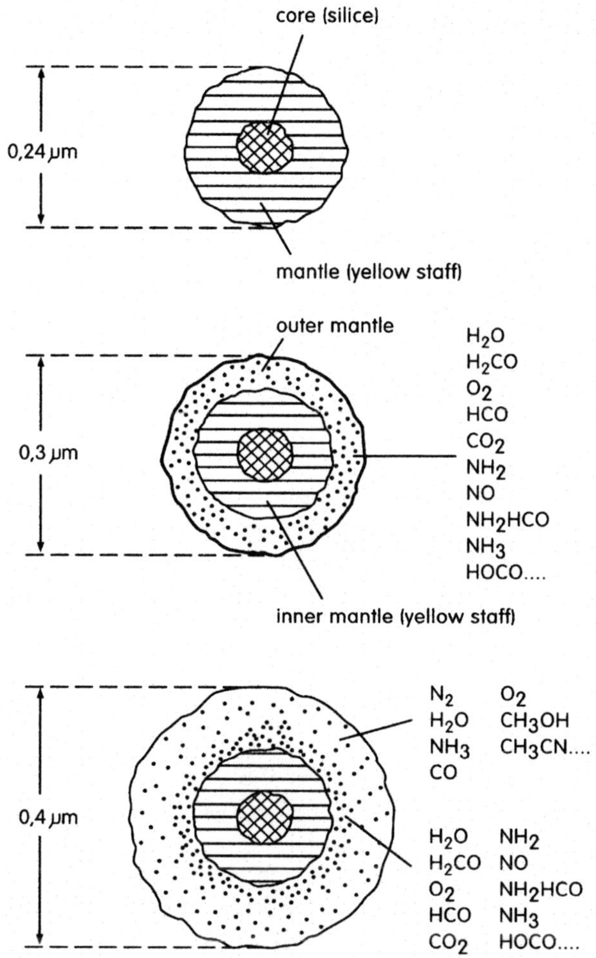

Fig. 7.1. The structure of the dust particles in dense interstellar clouds according to Greenberg (1984). The material of the internal mantle around the silicate core (yellow stuff) is a mixture of large, complex organic molecules. The external coat consisting of "dirty ice" contains simple organic compounds, ions, and freezed-in radicals.

so-called planetesimals). Stellesimals agglomerate into primitive stars by incorporating the material of the cloud around them. Therefore the primitive star is surrounded by a void.

Planet	Density (g/cm^3)
Mercury	5.44
Venus	5.23
Earth	5.52
Mars	3.95
Jupiter	1.34
Saturn	0.70
Uranus	1.58
Neptune	1.71
Pluto	1.00

Table 7.4. Density distribution of the planets.

The gravitational force of the primitive star attracts also the farther material of the cloud that accelerates toward the primitive star. However, when the speed of this material reaches a critical value of v_c, a very strong interaction takes place between the material and the magnetized plasma, and the fall of the particle toward the primitive star stops. The critical speed for the most frequent elements may be calculated from the equation:

$$\frac{1}{2}mv_c^2 = eV_{ion} \quad [7.1]$$

where m is the mass of the atom and V_{ion} is the ionization potential. This phenomenon of a critical speed has been verified by a thermonuclear experimental technique. In this way, material around the primitive star is arranged in a band structure. Inside a band, the phenomenon may be repeated; here the role of "primitive star" is played by the primitive planet around which the material ordered in a band structure is partly condensed into moons and partly stabilized as rings.

The Primitive Atmosphere

If our solar system would have been formed according to the previously discussed process and from the particles assumed by Greenberg (1984), it would be quite easy to reconstruct the composition of the primitive atmosphere. However, the high relative frequency of occurrence of Fe and other heavy elements in the solar system, as well as the presence of radioactive materials, suggests that the material of the solar system (or a major part of it) has not been formed according to the mechanism proposed by Greenberg, but originates from the explosion of a supernova. We do not have any information as yet concerning the structure, particle size distribution, composition, and fate of the materials originating from the explosions of supernovas. Thus, though there have been several quantitative models concerning the time course of the genesis of the solar system, we can only guess at the processes resulting in the primitive atmosphere of the Earth.

The foremost concept concerning the formation of the primitive atmosphere which even today has a great impact is that proposed by Urey (1952a, b, 1962). According to Urey, Earth agglomerated from meteorite like bodies (planetesimals containing iron, carbon, water, and hydrogen).

Because of the heat of agglomeration, volatiles left the planetesimals and, on reacting with each other, formed methane and ammonia as well as water and hydrogen. Thus, according to Urey, the main components of the primitive atmosphere were H_2, H_2O, CH_4, and NH_3. This gas mixture is of a strongly reducing nature. Unfortunately, we do not have reliable data on the composition of planetesimals. The composition of meteorites is not authentic from this point of view as they have been revolving for billions of years in a space from which the material of the clouds has already agglomerated into planets; their composition may well have been similar to that of comets.

In 1955, Rubey published a different concept for the origin and composition of the primitive atmosphere. According to this theory, the atmosphere developed slowly from gases liberated by volcanic activity from the inner part of the Earth. Correspondingly, at first H_2O, CO_2, Cl_2, N_2, and S were contained in the atmosphere, but the water was later condensed to form oceans. CO_2 was mainly bound in rocks forming carbonates while sulfur was bound in the form of sulfides and other sulfur-containing compounds. The atmosphere thus formed would be of a neutral or a slightly reducing nature.

Other published models may be regarded as variants of these two basic models except perhaps the concept that the atmosphere originates from the materials of comets colliding with Earth (Oró, 1961). The shortcoming of the details in the basic models results in very large deviations between calculations concerning the composition of the primitive atmosphere. Views agree insofar as the primitive atmosphere contained gases composed of biogenic elements and that its nature was definitely not oxidative. However, there are great differences as to the strength of its reducing character. According to Sagan and Mullen (1972), the content of the primitive atmosphere amounted to several parts per million, whereas Hart (1978, 1979) stated that it made up 25% of the total mass. Hart supposed a strongly reducing atmosphere, the composition of which is $2x10^{21}$ g CH_4, $4x10^{21}$ g NH_3, $6x10^{21}$ g CO, and $3x10^{21}$ g H_2S, or any combination corresponding to this composition.

It is interesting that the literature does not deal with the fact that the amount of carbon bound in carbonate rocks exceeds that found on Earth in other forms by orders of magnitude. If the atmosphere has not been formed continuously as the result of volcanic activity, the primitive atmosphere should have contained all the carbon now bound as carbonate. In this case, the atmospheric pressure on primitive Earth would have been about 200 bars at the surface. This is supported by the fact that on Venus, where it is probable that there are no carbonate rocks, the pressure at the surface is now about 100 bars, even though Venus is smaller than Earth.

Neither does the literature deal with the fact that, according to experiments concerning chemical evolution (to be discussed later), as a result of chemical processes taking place in the primitive atmosphere, water-soluble compounds are formed that, upon dissolution in water, cause a dark brown color not only in primitive oceans, but also in the condensed water droplets contained in clouds. Obviously, this must have basically changed the absorption relations of the radiation of the Sun, as well as the albedo of the whole of the Earth.

Finally, there is no agreement concerning the question as to whether Earth lived through a glowing state at all and that the separation of the iron core and the mantle occurred during the process of the formation of Earth or later. As a consequence, there is also a great uncertainty as to the temperature relations on primitive Earth. While the luminosity of the Sun was supposedly 20–30% lower than it is today, it is assumed that the intensity of its far-UV radiation could have been up to a hundred times higher (Canuto et al., 1983). It seems from several indications that the temperature of the Earth has not changed significantly during the past 4 billion years or more and this is explained by the greenhouse effect (Kuhn and Kasting, 1983; Owen, Cess, and Ramanthan, 1979; Rossow, Henderson—Sellers, and Weinreich, 1982; Sagan and Mullen, 1972).

There are a great number of papers that try to explain the composition of the primitive atmosphere via quantitative investigation of the kinetics of the reactions systems consisting of elementary photochemical reactions and other coupled chemical reactions (Ferris and Nicodem, 1972; Kuhn and Atreya, 1979; Lasaga, Holland, and Dwyer, 1971; Levine, 1982; Levine, Augustsson, and Natarajan, 1982; McGovern, 1969; Walker, 1977, 1978). Unfortunately, the numeric results of these papers, although in themselves very valuable works, cannot be accepted as reliable because of the uncertainties listed previously. What we can by all means accept is that:

1. Elements necessary to biogenesis were present in the primitive atmosphere in the form of simple compounds;

2. The energy needed for chemical synthesis was also provided in the form of radiation;

3. The environment was reducing, but to an unknown extent;

4. Water in great amounts and in liquid form has been present on Earth for as long as at least 4 billion years.

Chemical Evolution

Introduction

The overall name for a series of events the result of which is the formation of complex organic compounds, polymers, and microscopic structures from simple, C-, N-, O-, and H-containing molecules is **chemical evolution**. The total collection of processes that result in the appearance of living systems by utilizing the former products as raw materials, and from there lead to the simplest prokaryotes possessing enzymatic regulation and DNA-genes, will be called **prebiotic evolution**. Biological evolution leading to the formation of the living world starts with the appearance of the most primitive prokaryotes.

The conscious examination of chemical evolution started with the experiments of Stanley Miller who, advised by Urey, synthesized solutions containing amino acids and other organic acids by ultraviolet irradiation of, and electrical discharge in, gas mixtures consisting of H, C, O, N, and H_2O (Miller, 1953, 1955, 1957; Miller and Urey, 1959). Preceding those of Miller, experiments were carried out sporadically that would not be regarded as prebiotic, e.g., in 1898 Berthelot obtained carbohydrates by the irradiation of CO_2 and H_2O, in 1913 Loeb synthesized glycine from CO, NH, and H_2O by electrical discharge, whereas in 1953 Garrison succeeded in preparing formic acid and formaldehyde by the α-irradiation of a mixture of CO_2 and water (Buvet, 1974). But the experiments of Miller started an avalanche, and the abundance of experiments concerning chemical evolution nowadays is very difficult to survey. Thus, in what follows, the experimental results are surveyed in three groups separately, according to three different viewpoints (energy source, initial compounds, products) without a claim to completeness.

Energy source

The main components of the different primitive atmosphere models (CH_4 , CO, CO_2, N_2, NH_3, etc.) could not be the immediate starting materials of the processes of chemical evolution as they are stable, non-reactive compounds under the given conditions. For the triggering of the processes of chemical evolution, they should be transformed into very reactive compounds with high energy contents by the uptake of energy. In Miller's original experiment, electrical discharge served as the energy source. In this case the question arises as to what is the effective component in the process: the UV radiation generated by the discharge, the thermal effect, or the electrical energy itself?

By the application of thermal energy, various authors have succeeded in carrying out prebiotic chemical reactions. Harada and Fox (1964) showed the formation of amino acids in a gas mixture containing CH_4, NH_3, and H_2O in the presence of silicates. The same results were obtained by Oró at 1300 °C without silicates (Buvet, 1974). The effect of temperature itself has been studied by Lawless and Boynton; they vaporized a similar mixture and let it flow through a hot tube filled with quartz sand; thereupon the gases were absorbed by an ammonia solution and kept at 75 °C for 35 hours. Following evaporation, the hydrolysate was studied by an amino acid analyzer and gas chromatograph. At a tube temperature of 1060 °C, α-alanine, glycine, ß-alanine, succinic acid, ß-amino-n-butyric acid, and aspartic acid were found. At 980 °C, the three latter components were absent, and at 930 °C only α-alanine and glycine were present. At 1060 °C, 96% of the product was ß-alanine, whereas at 930 °C, 96% was glycine. At lower temperatures, numerous researchers have proved the presence of chemical evolution process without external energy sources, but in these experiments compounds of a high energy content were present among the starting materials, such as cyanide (e.g., Lowe, Rees, and Markham, 1963; Matthews and Moser, 1966; Oró and Kimball, 1961, 1962), formaldehyde (e.g., Fox and Windsor, 1970; Kenyon and Nissenbaum, 1976), or amino acids (e.g., Rohlfing, 1976).

Ultraviolet radiation. This is the most frequently used energy source of chemical evolution experiments. UV radiations of every wavelength that may be absorbed by the initial reaction components are effective.

Visible light. The energy of visible light is less than the energy needed for the activation of prebiotic starting molecules. However, Krishna Bahadur from India showed, in his extraordinarily interesting but not sufficiently appreciated work, that sunlight or even the light of a common electric bulb may be capable of starting the reactions if the required activation energy is decreased by applying catalysts (Bahadur, 1954; Bahadur, Ranganayaki, and Santamaria, 1958). In this way not only complex organic compounds, but also microscopic structures of very complex composition and functions may come into being (Bahadur, 1964, 1966, 1967, 1973–74; Bahadur and Ranganayaki, 1970, 1983; Bahadur et al., 1966). Similar experiences have been noted about the applicability of long-wave ultraviolet radiation, in which case H_2O can play the role of a primary photon acceptor (Sagan and Kare, 1971).

Among radioactive radiations, α-radiation was the first to be studied: in 1951, Garrison obtained formic acid and formaldehyde by the irradiation of wet CO gas, later malonic acid, hydroxysuccinic acid, succinic acid, and citric acid appeared upon the irradiation of acetic acid (Garrison et al., 1951). The effect of ß-irradiation (i.e., electron irradiation) on a mixture of methane, ammonia, and water was first investigated by Palm and Calvin (1962) and then by Oró (Calvin, 1969).

The effect of α-irradiation was mostly studied by I. and Z. Draganic (Draganic and Draganic, 1971, 1978; Draganic, Draganic, and Niketic, 1977; Niketic et al., 1982). The studies referred to here (and those not referred to) proved unambiguously that all of these three types of radioactive radiations are suitable energy sources for prebiotic processes.

Other radiations. On irradiating the aqueous solution of ammonium acetate by 1-MeV electrons, the formation of glycine and aspartic acid was observed (Hasselstorm, Henry, and Murr, 1957). Upon X-ray irradiation of a gas mixture consisting of hydrogen, methane, carbon dioxide, ammonia, nitrogen, and water, amines and amino acids are seen to be produced (Dose and Rajewsky, 1957). On the other hand, Hussmann (1968) observed the formation of nitrogen-containing heterocyclic compounds as the result of X-ray irradiation. Rössler (1984) studied the reactions of high-speed carbon atoms and established that organic molecules can also be formed upon the collision of these carbon atoms. This mechanism may well have a significant role in cosmic chemistry.

Electrical discharge can also serve as an energy source. In the original experiment of Miller, spark discharge was applied and his experiment was later repeated by numerous researchers using different gas mixtures. Glow discharge was also often used (e.g., Allen and Ponnamperuma, 1967; Harada and Suzuki,

1977; Raulin and Toupance, 1975). Finally, an interesting course was followed by Simionescu and Dénes in Romania; they generated cool plasma by the electrical discharge of a radiofrequency generator and studied the chemical evolution processes taking place in this plasma (Simionescu, Dénes, and Macuveanu, 1973; Simionescu, Dénes, and Totolin, 1981).

Initial compounds

From the viewpoint of the initial compounds, experiments on chemical evolution may be classified into two large groups. In the first group, the starting mixture consists of stable, gaseous compounds (e.g., CH_4, C_2H_6, CO, CO_2, N_2, NH_3, H_2, H_2O, H_2S) or a corresponding solution (containing, e.g., CO_3^- or NH_4^+ ions), whereas in the other group are active, aggressive compounds (e.g., HCOOH, HCHO, HCN, HCCCH). In the first group, the direct communication of energy is essential, whereas in the second one, the majority of the reactions take place even at ambient temperatures. The difference between the two groups is, however, mostly only virtual: it has been revealed that the primary intermediates arising upon the irradiation of members of the first group are members of the second one; thus the reaction products of experiments with components of the first group are really also produced from compounds belonging to the second group. Among the compounds in the second group, formaldehyde and hydrogen cyanide are of outstanding importance, the vast majority of the products of chemical evolution processes being synthesized from these two compounds.

The most important conclusions of chemical evolution experiments carried out by different researchers as to the initial materials of these processes may be summarized as follows.

1. Chemical evolution processes do not take place in oxidizing media (except for some special cases, e.g., in the presence of catalysts). This fact has been explained by Buvet, who showed that the free enthalpies of formation for compounds of biological importance are roughly one order of magnitude higher in a $CO_2 - H_2O - O_2$ mixture than in a mixture $CH_4 - H_2O - NH_3 - H_2$ mixture (Buvet, 1974).
2. The composition of the initial mixture can be arbitrarily varied in a very wide range. If the mixture consists of the simplest C-, O-, N-, and H-containing compounds, and the energy communicated is capable of activating several components of the system by being absorbed, chemical evolution processes start. Limitations are mainly due to the ratio of the

oxygen-containing or non-hydrogen-containing compounds. Miller (1983) showed in one of his latest experiments that the H_2/CH_4 ratio and the presence or absence of NH_3 do not influence the formation of amino acids, but if the H_2/CO or H_2/CO_2 ratio falls below 1, then in spark discharge experiments the yield decreases drastically. Glycine is the only amino acid produced in a CO and CO_2 atmosphere. Hattori (1983) obtained similar results with high-frequency electrical discharge experiments studying mixtures with different molar ratios. If the ratio of CO_2 exceeded that of CH_4, the yield decreased; on applying only CH_4 a dark, insoluble substance was formed in which, after hydrolysis, only very small amounts of amino acids could be detected.

3. Under reducing conditions, the chemical evolution processes resulting in the production of complex organic compounds take place very fast – in hours, days, or weeks. These time spans may be regarded as instantaneous within the geological time scale.
4. From gases that are insoluble or only soluble to a very small extent in water (CH_4, O_2, H_2, N_2), water-soluble compounds are formed at a high rate. Thus the majority of the processes take place in the presence of a liquid, aqueous phase. As the processes progress, more and more polymerized, insoluble substances are precipitated (as we shall see later, mainly in the form of microscopic structures of biological importance).

Low-molecular-weight products

Amino acids were the first detected, most frequently appearing, and most investigated products. They are mainly produced by Strecker synthesis from primary products such as aldehydes and HCN, but they may also be formed by the hydrolysis of cyan polymers produced under dehydrated conditions (Matthews and Moser, 1966, 1967). In different prebiotic experiments, only 4–14 amino acids are found that are constituents of proteins, but there are also non-protein-forming amino acids present (e.g., β-alanine, α-, β-, and γ-aminobutyric acid, α- and β-aminoisobutyric acid, isovaline, norvaline, norleucine). Cyclic amino acids are generally not found, though under certain circumstances they can also be formed (Friedmann and Miller, 1969). Neither can sulfur-containing amino acids be formed, as among the initial compounds there are usually no sulfur-containing substances. However, upon the UV-irradiation of ammonium thiocyanate, methionine has been found among the products (Steinmann, Smith, and Silver, 1968).

Mono- and dicarboxylic acids are formed with a similar frequency and with as high a variety as the amino acids. In his first experiments, Miller could detect, for example, formic acid, glycolic acid, lactic acid, acetic acid, propionic acid, α-hydroxybutyric acid and succinic acid in addition to amino acids. Yuen and co-workers observed the formation of 17 monocarboxylic acids in spark discharge experiments, and also measured their amounts. The number of carbon atoms in these acids varied between 2 and 7, whereas higher carboxylic acids were produced only in smaller quantities (Yuen, Lawless, and Edelson, 1981).

Carbohydrates. In spite of the early discovery of their spontaneous formation (Butlerov, 1861), the autocatalytic nature of which was disclosed by Neuberg as early as 1902 and investigated in detail by Langenbeck in 1942, and in spite of the fact that their prebiotic significance was strongly emphasized in numerous works of Decker (Decker 1973, 1974, 1979; Decker and Speidel, 1972; Decker and Heidmann, 1978; as well as in works by other authors (Cairn-Smith, Ingram, and Walker, 1972; Cairn-Smith and Walker, 1974; Gánti, 1978b, c; Quayle and Ferenci, 1978), their role in chemical evolution is still not fully understood and experiments on this subject are sparse (Decker, 1974; Degani and Halmann, 1967, 1971; Gabel and Ponnamperuma, 1967; Reid and Orgel, 1967). Knowledge of the mechanism and kinetics of the formose reaction does not in fact stem from chemical evolution experiments (see the comprehensive works of Mizuno et al., 1970–72; Mizuno and Weiss, 1974).

Sugar formation from formaldehyde, the so-called formose reaction, takes place in basic medium. The first step of the reaction is an aldol condensation: from two molecules of formaldehyde one molecule of glycolaldehyde is formed. This reaction step is very slow. The subsequent steps are also further aldol condensations and intramolecular rearrangements, but they proceed at a much faster pace. Thus glyceraldehyde, 1,3-dihydroxy-2-propanone, then aldoses and ketoses of 4–6 carbon atoms are produced together with ketoses of 7 or 8 carbon atoms. On applying carbonate apatite instead of basic media, sugars of 5 or 6 carbon atoms are formed from 0.05 M formaldehyde with a yield of above 40%, if they are boiled (Reid and Orgel, 1967), whereas 80% of the formaldehyde is transformed into sugars on using alumina as a catalyst (Gabel and Ponnamperuma, 1967). The latter authors also successfully used kaolinite and illite as catalysts.

In a flow-through reactor, Decker studied the reaction of a 0.158 M formaldehyde solution containing 0.0825 M NaOH and 0.036 M calcium acetate at 57 °C, at a flow-through time of 3 min. He obtained a conversion of 97%, the products detected being treose, eritrose, eritrulose, ribose, ribulose, arabinose, lyxose, xylose, xylulose, arabinulose, fructose, and sorbose. It is worth noting that the same products were also obtained in the absence of Ca^{++} ions, at a flow-through time of 3 min and at 50 °C, if a mixture of glycolaldehyde and glyceraldehyde was used as initial material. A significantly more complex reaction network than that described earlier is shown, for the illustration of the formose reaction, in Fig. 7.2. It can be clearly seen that though the formose cycle is much more complex than the scheme shown earlier (which contains only the inner cycle of this reaction network), the autocatalytic nature is preserved. Moreover, there is a feedback from the sugars of 5 or 6 carbon atoms, which makes autocatalysis possible also here.

Sugar phosphates. In spite of their central role in the metabolism of living organisms, their prebiotic formation and role has been little studied. It is only known that in sodium phosphate solutions, in the presence of cyan derivatives (e.g., cyanogen), glycose-1-phosphate, glycose-6-phosphate, and a small amount of glycose-1,6-diphosphate are formed from glycose. The authors mention that the intermediates of glycolysis have also been formed in their experiments (Degani and Halmann, 1967, 1971, 1972).

Nucleotide bases. The fundamental route for the synthesis of purine bases was discovered by Oró and Kimball (Oró, 1960; Oró and Kimball, 1961, 1962). On letting an ammonia-containing solution of hydrogen cyanide stand, adenine, aspartic acid, alanine, formamide, formamidine, glycine, glycine-amide, 4-aminoimidazole-5-carboxamide, and 4-aminoimidazole-5-carboxyimidine have all been detected in the reaction mixture. According to the reaction pathway suggested by the authors, adenine molecules are built up of five HCN molecules while the NH_3 is consumed and formed, i.e., NH_3 plays only the role of catalyst in the system (Fig. 7.3). According to Orgel and co-workers, in the presence of guanidine or urea the end product is guanine or xanthine, respectively (Fig. 7.4). The reaction has been studied extensively (Sanchez, Ferris, and Orgel, 1966a, b, 1967, 1968; Ferris et al., 1974; Mizutani et al., 1975; Schwartz, Toosten, and Voet, 1982; etc.) and yet the reaction path is still not fully clarified. There are probably several alternative pathways possible. The statement that formaldehyde and other carbonyl compounds accelerate significantly the oligomerization of HCN to adenine and other products is very remarkable (Schwartz et al., 1982; Schwartz, 1983). The effect is probably exerted through the additional product of formaldehyde and HCN, the glycolonitrile.

Ferris and co-workers referred first to the possibility of the abiogenic synthesis of pyrimidine. In addition to adenine and guanine, they obtained also cytosine and uracil from CH_4 in electrical discharge experiments (Ferris, Sanchez, and Orgel, 1968), or else from the cyanoethylene intermediate from which first cytosine is formed, the hydrolysis of which then results in uracil. In the 1970s, numerous works dealt with the possible synthesis of pyrimidines (Sanchez and Orgel, 1970; Ferris et al., 1974; Choughuley et al., 1978; Schwartz and Chittenden, 1978). Especially interesting for us are the investigations of Orgel et al., who have shown that the pyrimidine base may be directly synthesized on ribose from cyanamide and cianoacetylene.

Phosphorus-containing compounds play a very important role in living organisms. In the 1960s, numerous unsuccessful attempts were made especially to prove the prebiotic synthesis of ATP. Sporadic data in the literature concerning the synthesis of ATP could not be reproduced, with a single exception: the work of Akaboshi and Kawai (1971). However, this successful experiment cannot be regarded as typical: the authors observed the *in statu nascendi* building of atoms originating from the decomposition of atoms, as a result of neutron irradiation of adenosine, AMP, and ADP obtaining thereby 2'-AMP, 3'-AMP together with ADP and ATP.

The role of a phosphorylating agent may be played, under abiotic conditions, by much simpler compounds than ATP. First, pyrophosphate and linear polyphosphates were considered (Schwartz and Ponnamperuma, 1968), the prebiotic formation of which was studied by several authors (Osterberg and Orgel, 1972; Weber, 1981, 1982), as well as inorganic thiophosphate (Slabaugh, Harvey, and Nagyvári, 1974). In a gas mixture containing phosphine, a multitude of phosphoruscompounds have been formed in electrical discharge experiments (Rabinovitz et al., 1969).

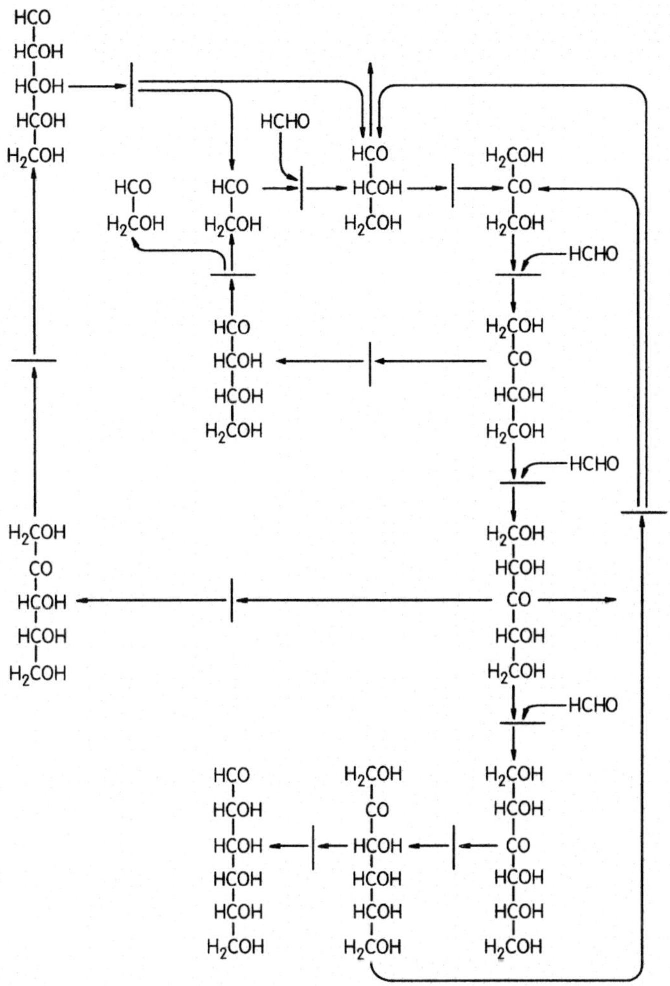

Fig. 7.2. The formose cycle in a more detailed representation. It is apparent that a stoichiometric feedback making autocatalysis possible exists also in pentoses and hexoses.

A very plausible route of phosphorylation has been found by Halmann and co-workers (Degani and Halmann, 1967, 1971, 1972; Halmann, Sanchez, and Orgel, 1969; Halmann, 1975), who have shown that in the presence of orthophosphate and cyan derivatives sugars transform into sugar phosphates. Under similar conditions, Rao and co-workers found the formation of phosphatidylcholine (Rao, Eichberg, and Oró, 1982). It can be supposed that the phosphorylating agent in the reaction is the cyanovinylphosphate found to be formed (Ferris, 1968; Ferris, Goldstein, and Heaulien, 1970). Carbamylphosphate is a similarly good phosphorylating agent that can also be formed under prebiotic conditions (Lohrmann and Orgel, 1971; Schwartz, 1972; Osterberg and Orgel, 1972; Saygin and Ellmaurer, 1983). In the presence of Ca^{2+} ions, ADP could be phosphorylated with carbamylphosphate, yielding 18% of ATP! Regarding reaction in the presence of Ba^{2+}, Mn^{2+}, Cu^{2+}, or Co^{2+} ions, AMP can be phosphorylated to ADP under similar conditions. It is very important to note that, in the presence of Al^{3+} or Be^{3+} ions at 37 °C, acetate transforms to acetylphosphate in 30 minutes with a conversion ratio of 30! (at a pH of 4.5) as shown below:

$$CAP^{2-} + CH_3COO^- \xrightarrow{Be^{2+}} CH_3COOPO_3^{2-} + \text{carbamate}$$

(Saygin and Ellmaurer, 1983).

Other low-molecular-weight products. Data in the literature concerning the spontaneous formation of other biochemically important compounds are very scarce. There are opinions stating that, from primitive methane, so many higher hydrocarbons could be polymerized on UV irradiation that they would cover the Earth to a thickness of 1–10 m (Lasage, Holland, and Dwyer, 1971). From hydrocarbons, in turn, fatty acids could be formed upon irradiation (McKusick, Mochel, and Stacey, 1960). C_6–C_{12} fatty acids could also form directly from methane in electrical discharge, but the C_6–C_{12} fatty acids thus obtained are mainly branched ones (Allen and Ponnamperuma, 1967). Glycerides could also be formed (Oró, 1963; Sherwood and Oró, 1976; Deamer and Oró, 1980).

The formation of lipidlike compounds and structures was observed as effected by cold plasma (Simionescu, Dénes, and Totolin, 1983). After a chloroform extraction, in the aqueous phase, mainly acrylic acid amide could be detected in the interphase as its polymerization product:

$$(CH_3\text{—}(CH_2)_{n1}\text{—}(\underset{\displaystyle CONH_2}{\underset{|}{CH_2\text{—}CH}})_{nz}\text{—}\underset{\displaystyle CONH_2}{\underset{|}{CH_2\text{—}CH_2}})$$

and, in the chloroform phase, olefins containing one double bond.

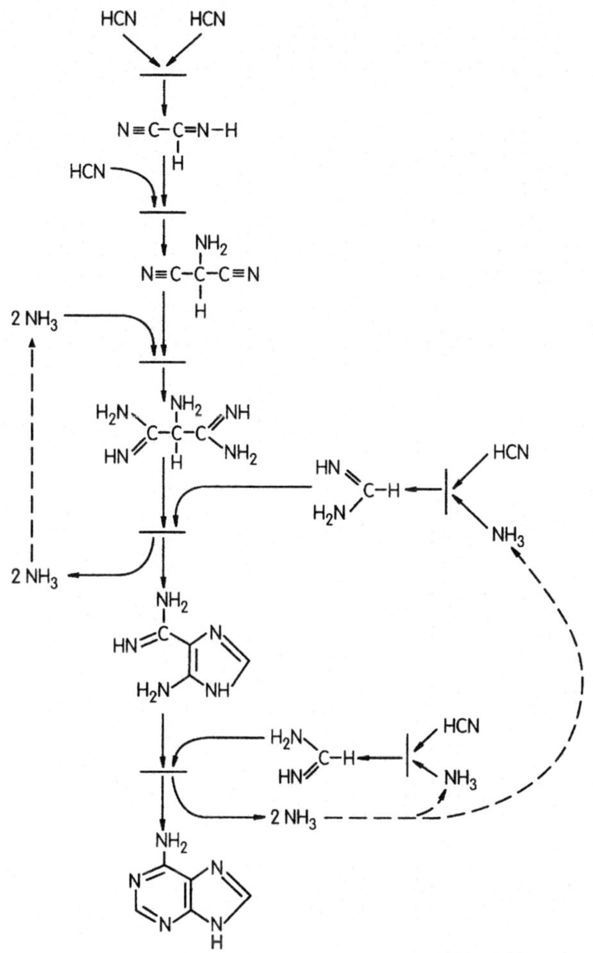

7.3. The prebiotic synthesis of adenine according to Oró and Kimball.

The prebiotic intermediate of isoprenoids could be apiose (Decker and Schveer, 1983) which may be considered the prebiotic analogue of isopentenylphosphate.

Upon irradiation of ammonium cyanide, urea and guanidine could be obtained (Lohrmann, 1972), whereas upon irradiating aqueous formaldehyde, glyoxal and malonic aldehyde are also formed (Halmann and Bloch, 1979). A reaction between malonic aldehyde and urea may lead to the synthesis of pyrimidines.

On electrical discharge in CH_4/H_2O and $CH_4/N_2/H_2O$ mixtures, propionaldehyde is formed that – upon reacting with cyanoacetaldehyde and ammonia – provides nicotinonitrile, the hydrolysis of which in turn forms nicotinamide and nicotinic acid (Dowler, Fuller, and Orgel, 1970). The formation of other nicotinic acid derivatives has been observed by other authors (Ferris, Kuder, and Catalaro, 1969; Friedmann et al., 1971).

It is very important that, among the products of chemical evolution, imidazole also appears (Ferris and Kuder, 1970; Oró et al., 1983), as imidazole catalyzes the condensation of mononucletoides (Ybanez, Kimball, and Oró, 1971a). Porphin derivatives were also detected (Szutka, 1964; Hodgson and Baker, 1967; Hodgson and Ponnamperuma, 1968; Hayatsu et al., 1972).

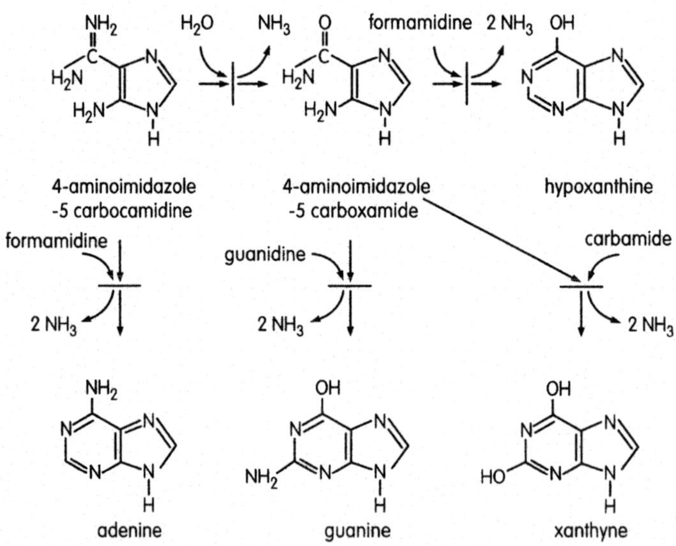

Fig. 7.4. Guanine and xanthine may also be formed from the products of the polymerization of HCN and cyanoacetylene (Sanchez and Orgel, 1970; Orgel and Lohrmann, 1974). The synthesis route suggested by them is shown in Fig. 7.5.

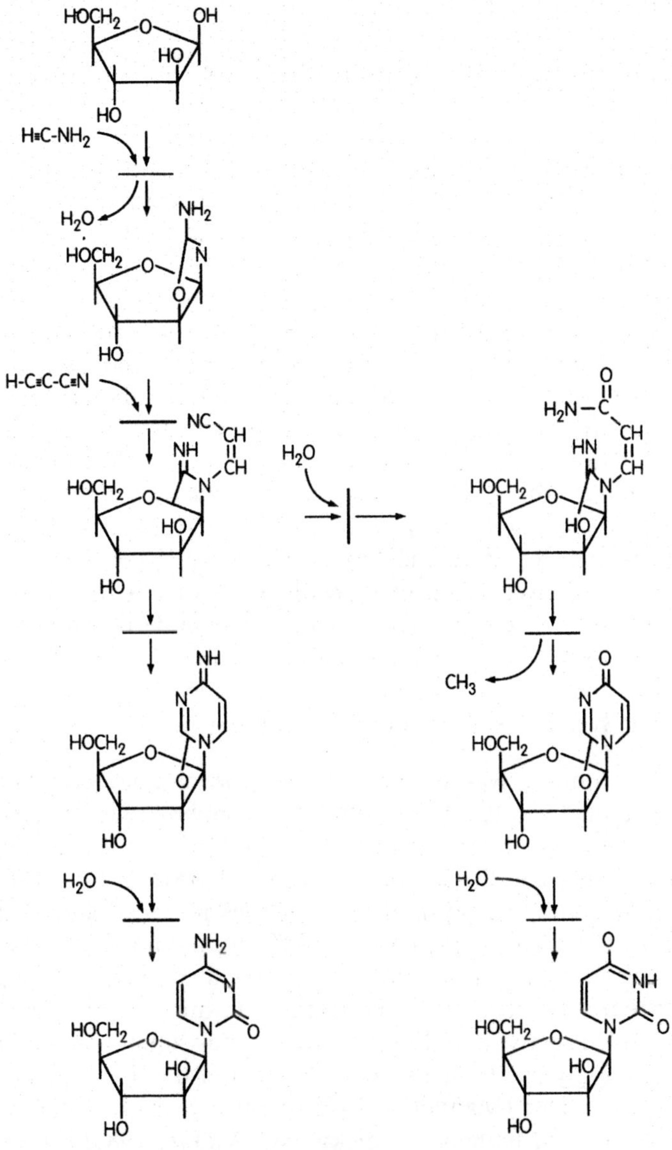

Fig. 7.5. Prebiotic synthesis of pyrimidines on a ribose molecule.

Abiotic formation of proteinoids

In 1959, Fox and co-workers observed that mixtures of amino acids, especially mixtures containing great amounts of basic amino acids, polymerize (upon dry heating) to proteinlike compounds, to so-called proteinoids. The formation of thermal proteinoids has been extensively studied since (Fox and Harada, 1960; Fox, 1964, 1965, 1973a, b, c), and it has been established not only that heteropolycondensation occurs, resulting in high-molecular-weight (molecular weights up to 300,000) proteinoid macromolecules, but also that microspheres surrounded by a membrane and having a diameter of several micrometers are of proteinoids. It has been shown that neutral proteinoids may be synthesized (Fox and Waehneldt, 1968) and that the synthesis of thermal proteinoids at less than 100 °C is possible (Rohlfing, 1976). It has been observed that the composition of proteinoids does not fully reflect the amino-acid composition of the starting mixture, thus the formation of the amino-acid sequence is not fully random and the proteinoids may have some primitive enzymatic (catalytic) activity. Some hydrolase (Hardenbeck, Krampitz, and Wulf, 1968), phosphatase (Oshima, 1968), and other activities (Rohlfing and Fox, 1969; Dose, 1974; Fox, 1973c, 1975) have been observed. According to more detailed studies, however, not only α- but, in the majority of cases, also β-, γ-, and ε-peptide bonds are to be found (Temussi et al., 1976). In thermal lysine co-polymers α- and ß-peptide bonds occur upon the coupling of lysine and aspartic acid, whereas α- and ε-bonds are formed upon lysine–lysine coupling (Fox and Suzuki, 1976).

A totally different possibility was pointed out for the abiotic formation of proteins by Matthews and Moser (1966, 1967), who polymerized water-free HCN in the presence of a small amount of NH_3, then dissolved the precipitate formed, fractionated it by gel filtration, and obtained products of different molecular weights. An acidic hydrolysis of the fractions provided amino acids. In the hydrolysates of the various fractions, 14 different amino acids could be detected, but in every fraction, excess glycine was found. In this case, peptide formation did not occur through the polymerization of amino acids, but rather HCN was polymerized directly to a linear polymer, polyaminomalononitrile, which transforms to polyglycine by the slow hydrolysis of the imino and cyano groups on contact with water (Matthews, 1975; Matthews and Moser, 1966, 1967; Matthews et al., 1977; Matthews, Ludicky, and Widing, 1983; Moser and Matthews, 1968; Moser, Clagett, and Matthews, 1968a, b). The first part of this process, i.e., the formation of cyan polymers, probably takes place also in interstellar clouds, on the surface of meteorites, etc. However, some doubt arises as to its role in prebiotic processes, as the peptides thus formed are resistant to the effect of prebiotic enzymes (Ferris, 1979; Ferris et al., 1973). At the same time, Niketic and co-workers showed that in aqueous phases, upon ionizing radiation, cleavable peptides are enzymatically formed from cyanides. The molecular weights of these polymers are between 1 and 20.000, and about 30% of the polymers are polypeptides (Niketic, Draganic, and Draganic, 1983).

Numerous authors studied the possibilities of the condensation of amino acids to peptides in aqueous solutions. It became obvious that peptides may also form in experiments using electrical discharge (Flores and Ponnamperuma, 1972; Simionescu et al., 1973). Clay minerals were also found to enhance the coupling of amino acids to peptides (Paecht-Horowitz, 1974; Paecht-Horowitz and Lahaw, 1977; Warden et al., 1974; Degens, Matheja, and Jackson, 1970; Paecht-Horowitz, Berger, and Katchalsky, 1970; White and Erickson, 1981; White, Macklin, and Kennedy, 1983), and it has also been shown that amino-acid adenilates enhance peptide formation both in the presence and absence of clay minerals (Paecht-Horowitz et al., 1970; Moiseeva, Pavlovskaya, and Sirku, 1983). Finally, similar effects have been observed in the presence of polymetaphosphates (Nooner and Oró, 1974) or pyrosulfuric acid (Dénes and Fox, 1976).

Abiogenic formation of nucleic acids

The formation of phosphodiesteric bonds under prebiotic conditions was first reported by Czech scientists, who heated uridine or 6-azauridine to 160 °C in the presence of inorganic phosphates, and obtained 3'–5' or 2'–5' dinucleoside phosphates and cyclic phosphates (Skoda and Moravek, 1966; Moravek and Skoda, 1967). In the presence of polyphosphoric acid, cytidylic acid was condensed by Schwartz and Fox in the same year (Schwartz and Fox, 1967).

Great progress was made by the realization that prebiotic nucleic acid formation probably took place not through nucleotide triphosphates but though nucleotide imidazolides. Thus, as early as 1968, Weimann and co-workers succeeded in proving that adenosine-5'-monophos-phorimidazolide was oligomerized spontaneously with adenosine derivatives to di- and trinucleotides, and that the process was significantly accelerated by the presence of polyuridylic acid, i.e., that in the reaction mixture a nonenzymatic template polymerization took place (Weimann et al., 1968). Simultaneously, the template polymerization of adenosine-5'-phosphate on poly-U was also observed, in the presence of carbodiimide as a condensation reagent (Sulston et al., 1968). In the same year, it was also shown that in the presence of the poly-U template D-adenosine-5'-phosphorimidazole reacts much faster with D-adenosine than with L-adenosine

(Schneider-Bernlohr et al., 1968). The condensation of deoxyribonucleotides was also observed in an aqueous solution at neutral pH, in the presence of imidazole (Ybanez et al., 1971a). It turned out later that cyanamide can also play the role of a condensation reagent (Ybanez, Kimball, and Oró, 1971b; Oró and Stephen- Sherwood, 1974).

These experiments have proved unambiguously that (1) the formation of a phosphodiesteric bond is possible under prebiotic conditions, and (2) for the polycondensation of ribonucleotides the template effect operates even in the absence of enzymes. At the same time, the experiments have also indicated that (1) polymerization results only in the formation of oligonucleotides (mostly $n < 8$) and (2) that in addition to the 3'–5' phosphodiesteric bonds present in nucleic acids, 2'–5' phosphodiesteric bonds are also formed, on an even larger scale. Therefore the observation of Orgel and co-workers, that Pb salts and Pb-containing minerals catalyze the formation of oligonucleotides from nucleotide-5'-phosphorimidazolides to an extent proportional to their solubility, is of great importance (Sawai and Orgel, 1975; Sawai, 1976; Sleeper and Orgel, 1979). They obtained $n > 5$ oligomers from ImpA with a yield of 18% under optimum conditions. In further experiments, this yield could be increased, in the presence of poly-U, to 35% (Sleeper, Lohrmann, and Orgel, 1979). Moreover, what is even more important is that 75% of the phosphodiesteric bonds formed were 3'–5' bonds, whereas this figure was only 10% in the uncatalyzed reaction.

The reaction can be catalyzed by other divalent cations as well. Mg^{2+} ions enhance mainly the formation of oligoadenylates with 2'–5' bonds on a poly-U template. On a poly-C template, both Pb^{2+} and Zn^{2+} ions enhance the formation of oligoguanylates from guanosine-5'-phosphorimidazole, but using Pb^{2+} mainly 2'–5' bonds are formed, whereas with Zn^{2+} the great majority are 3'–5' bonds (Lohrmann et al., 1981). At the same time Zn^{2+} inhibits the polymerization of ImpA on a poly-U template.

The effect of the properties of the template on the Pb^{2+}-catalyzed reaction has been investigated in detail by Sawai (1981). He found that if either Pb^{2+} or poly-U ($n<6$) is present alone in the reaction mixture, the formation of 2'–5' bonds is preferred. If $n>8$ poly-U is used as a template, mainly 3'–5' bonds are formed. The simultaneous presence of Pb^{2+} and poly-U strongly enhances the formation of 3'–5' phosphodiesteric bonds. Upon studying the effect of the length of the template, Sawai found that oligo-U with $n<6$ decreased the yield of oligo-A formed from ImpA, as ImpA gave heterooligonucleotides on reacting with oligo-U. On the other hand, oligo-A with $n>8$ unambiguously enhanced the formation of $n >5$ oligo-A.

Similar experiments have also been performed with oligo-C templates (Fakhrai, von Roode, and Orgel, 1981). It was found that a significant template effect is present even with n >2 oligo-C, and that this effect has its maximum at n=6. Short templates are capable of enhancing the formation of oligo-G longer than them from ImpG. In the absence of divalent metal ions, mixed bonds are formed, in the presence of Pb^{2+}, 2'–5' bonds, whereas with Zn^{2+}, 3'–5' bonds predominate.

Studies described so far applied poly-U (oligo-U) or poly-C (oligo-C) as templates, i.e., homopolymers of pyrimidine nucleotides. There is no description in the literature about a successful application of the homopolymers of purine nucleotides. This is a basic difficulty, because on the homopyrimidine strands homopurine strands are necessarily synthesized, and if these latter are not capable of functioning as a template, replication cannot be accomplished. In order to clarify this problem, three routes were followed. First, it was supposed that if the polymer is built not from the activated derivatives of mononucleotides, but from those of di- or trinucleotides, the process may also take place on a purine template. Activated di- and trinucleotides in fact transformed to oligomers on a poly-C template (Lohrmann et al., 1981), but it is not known whether the process could be carried out also on a purine template.

Very interesting results have been found in the second direction; Lohrmann investigated how the precipitate formed from the metal salts of mononucleotides influences the condensation of nucleotide-5'-phosphorimidazolides. He found that the precipitates of the metal salts studied (Pb^{2+}, Zn^{2+}, Mn^{2+}, Ca^{2+}, Al^{3+}, Fe^{3+}, La^{3+}) catalyze very strongly the condensation of various nucleotide imidazolides, corresponding to the base-pairing rule of Watson and Crick. Thus on the Pb^{2+} precipitate of uracil or thymine, adenosine and ImpA are condensed with an efficiency of 95%, and similar results were obtained with the condensation of G and ImpG on a C-containing matrix. The reaction also takes place on a matrix made of the salts of purine nucleotides, the condensation of ImpA and ImpG resulting mainly in 2'–5' bonds, whereas that of ImpC leading in 89% to the formation of 3'–5' bonds (on a Pb deoxyguanylate matrix, as the Pb guanylate matrix proved to be inefficient) (Lohrmann, 1982).

The third way, i.e., studying heteropolynucleotide templates, is the most difficult technically, but at the same time it is also the most promising one, as the formation of homopolymers under primitive conditions is most improbable. Orgel (1983) used various co-polymers containing excess C as a template, and found that, in the oligonucleotides formed, nucleotides satisfying the base-pairing rule were always present. On using CCGCC oligoribonucleotide as a template, the GGCGG oligonucleotide could be detected, among other products.

Joyce used cytosine-rich RNA co-polymers as templates for the condensation of mononucleotide-5'-phospho-2-methyl imidazolides. He showed that in the newly synthesized oligomers only those monomers appeared the complementaries of which were contained in the template. However, the efficiency of the template decreases rapidly if the amount of cytosine contained in it becomes smaller. Among the templates with low C contents, the more efficient are those in which the regions not containing cytosine are better distributed along the template (Joyce, 1983a; Joyce et al., 1984).

Formation of microscopic structures

The origin of the microscopic structures in living organisms was first derived from the effect of simple physicochemical forces; and noted by Oparin (1957) when he supposed that the cellular state developed from so-called coacervates. Under suitable conditions, the solutions of hydrophilic colloids do not coagulate but are separated into a colloid-rich and a (practically) colloid-free phase. The colloid-rich phase was named **coacervate** by Bungenberg de Jong, and he established that it consists of a great number of microscopic colloid spherules floating in an aqueous solution (Bungenberg de Jong, 1932). In his experiments, the diameter of these spherules varied between 2 and 670 μm.

Oparin found that the properties of these coacervate droplets are in many respects similar to those of protoplasm. Moreover, if the cell membrane is ruptured, cytoplasm is not mixed with water, but forms many coacervate dropletlike particles (Oparin, 1957). In coacervate droplets vacuola may be formed and the droplets are capable of absorbing the materials of their environment selectively. The interfaces of coacervate droplets (according to the knowledge of that time) seemed to be similar to the interface of cytoplasm; coacervate droplets of various substances showed multiple inclusions and appeared to model the structure of cell organelles. Based on all this, Oparin thought that the organic material formed in the primitive ocean agglomerated to coacervate droplets spontaneously and the evolution of these droplets led to the appearance of the cell in its present form. However, with the development of electron microscopy it became clear that the structure of coacervates cannot be brought into direct correlation with the microstructure of the cell. The coacervate hypothesis of Oparin is today only of a science-historical value: it directed the attention of researches to the possibility of a spontaneous formation of microscopic structures on a physicochemical basis.

The next step in the research of the formation of microstructures was made by Sidney Fox, who discovered that on the dissolution of the material obtained by heating amino acids, not only water-soluble proteinoids appear, but also tiny insoluble spheres with diameters of about 1 μm, and he called these "microspheres." The properties of these microspheres have been thoroughly investigated since (Fox, 1964, 1965, 1973a, b, c, 1974a, b; Fox and Dose, 1972; Fox, McCauley, and Wood, 1967; Fox, Lacey, and Nakashima, 1971; Waehneldt and Fox, 1968; Hsu, Broke, and Fox, 1971, etc.), and it has been established that in contrast to coacervates, which tend to aggregate and have a very broad size distribution, microspheres are of nearly identical size and are extraordinarily stable; they are surrounded by a double-layered membrane consisting of thermal proteinoids and having the nature of a two-dimensional liquid. Microspheres may be initiated to divide by changing the external conditions; in concentrated proteinoid solutions, buds are formed on them by the growth and subsequent separations of which (i.e., by budding) new microspheres may be formed. It has been shown that the membrane of microspheres has multiple catalytic activities and that at the membrane a "membrane potential" can be measured. The composition of the solution inside the microspheres differs from that of the external solution, i.e., the membrane has a selective permeability. The membrane of each microsphere is built of about 10^{10} proteinoid molecules. The basic proteinoids can interact with DNA and RNA, and thus microspheres containing nucleic acid may be formed, etc.

At about the same time as thermal proteinoid microspheres were discovered (1963), Krishna Bahadur and co-workers in India established that, if a sterile mixture of formaldehyde, ammonium phosphate, mineral salts of biological importance, and ammonium molybdate is exposed to the effect of sunlight, spherical microscopic structures are formed, which they called "Jeewanu". This is a Sanskrit word meaning "particles of life" (Bahadur, 1964, 1966a, 1967, 1973/74; Bahadur and Ranganayaki, 1983). These spherules are also separated from their surroundings by membranes which are semipermeable. They are in many respects very similar to the proteinoid microspheres but, in contrast to those, they have a very complex composition. Besides amino acids, phospholipids and carbohydrates have also been detected in them. Like proteinoid microspheres, the Jeewanu also possesses multiple catalytic activities. The Jeewanu formation is one of the most interesting and promising experiments in the research into the spontaneous formation of living structures, and one that is unjustly neglected in the literature. At a later point, in connection with the synthesis of living systems, it will be dealt with in more detail.

On UV irradiation, microspheres were obtained by Fraser and Folsome. Their starting mixture contained N_2, CH_4, CO_2, and liquid water. They also established that the rate of microstructure formation corresponded to autocatalytic kinetics (Fraser and Folsome, 1975; Folsome, 1976, 1977; Folsome and Brittain, 1981).

Kenyon and Nissenbaum even prepared two different kinds of microstructures: from the solution of amino acids and sugars, so-called melanoidin microspheres were formed, whereas on simply letting a formaldehyde solution of ammonium cyanide and other ammonium salts stand, aldocyanine microspheres were obtained (Kenyon and Nissenbaum, 1976; Pollock and Heiderer, 1979). In formaldehyde-containing amino-acid solutions corresponding in salt composition to modified seawater, Egami and co-workers observed the formation of microspheres that they called **marigranules** and **marisomes**.

Simionescu and Dénes observed the formation of organic microstructures in a quite different way. They let a mixture of CH_4 and NH_3 flow through an ice surface cooled to -60 °C, in the presence of an RF glow discharge. After the reaction, when the ice was thawed, stable spheres of diameter 10–50 μm were obtained. Upon heating to 80 °C or UV irradiation, these spherules are partly dissolved, but when the original conditions are restored, they are formed again (Simionescu et al., 1981, 1983). However, from their reports it seems that these microstructures are rather of an emulsion droplet character than microspheres surrounded by a membrane, as are the Jeewanu or the proteinoid microspheres.

VIII. Biogenesis

Models of Biogenesis

Experimental models

No kind of random group of an arbitrary multitude of biologically important compounds is a living system. Therefore whatever multitude of compounds has been formed during chemical evolution in whatever concentration, this does not yet mean the formation of life, only the accumulation of raw materials needed for the genesis of life. Living systems are program-controlled proliferating fluid chemical automata with a defined internal organization. Thus the key question in the genesis of life is whether the organic compounds produced in chemical evolution could combine to such functioning, self-regulated, program-controlled, proliferating systems. Prebiotic evolution begins with this combination process and it ends with the development of hereditary information enclosed in the DNA genes and the regulation by protein enzymes coupled to them.

Some researchers try to discover the processes of prebiotic evolution by experiments, some by qualitative theories, and others discuss the topic on a more or less quantitative basis, by means of quantitative models. Theoretical attempts are, of course, based on experiments, whereas experimental works lack a fundamental theoretical basis and they are based on either random discoveries or experimental considerations.

The pioneer of experimental models was Oparin, who considered the key step in the formation of life the production of coacervates (Oparin, 1957). As was mentioned in the previous chapter, numerous similarities could be observed between the properties of coacervates and those of cells. However, because no plausible mechanism of the formation of life could be outlined on the basis of the coacervate hypothesis, it is today only of science-historical importance.

Sydney Fox considers the formation of thermal proteinoid microspheres as a plausible model for the origin of life (Fox, 1965, 1973a, b, c). To support this hypothesis, Fox and co-workers revealed a number of interesting properties of proteinoid microspheres, first of all various catalytic activities ("enzyme activities"), morphological characteristics, electrical properties (Ishima, Przybylski, and Fox, 1981; Przybylski et al., 1982; Przybylski and Fox, 1984; etc.). In spite of all this it seems that the formation of thermal proteinoid microspheres cannot be regarded as a real mechanism for the origin of life, first of all because the growth of the spheres is not the result of an internal metabolic activity of the system, but it occurs as a consequence of the incorporation of the components of the external proteinoid solution into the membrane. The microspheres of Fox are good models for the formation of closed membrane surfaces ensuring compartmentalization, but they do not model either the spontaneous formation and operation of self-reproducing fluid chemical machines, or the formation of the self-reproductive program, i.e., that of the hereditary material.

Krishna Bahadur, the discoverer of Jeewanus considers the formation of these microspheres the model for the origin of life (Bahadur, 1964, 1966, 1967; Bahadur and Ranganayaki, 1983). In fact, it seems from his experiments that Jeewanus may be the models not only of compartmentalization but also of internal chemical machines, because they were shown to be capable of multiple, complex synthetic activities. In all probability, the microspheres of Fraser and Folsome possess similar properties (Folsome, 1976, 1977), as their formation follows exponential kinetics (Fraser and Folsome, 1975), which refers to an autocatalytic behavior, and they are capable of the photosynthetic incorporation of CO_2 (Folsome and Brittain, 1981), which allows for the operation of an internal metabolic network. There are no data concerning either the microspheres of Bahadur or those of Folsome that would indicate the presence of an internal program. A much detailed analysis of these experiments will be made later concerning the synthesis of living systems.

Qualitative theoretical models

Numerous trains of thoughts are devoted in the literature to certain questions of prebiotic evolution. In particular many speculations and assumptions have been published on the origin of the genetic code and of the protein-synthesizing apparatus (Pattee, 1961; Jukes, 1965, 1974, 1981; Bloom, 1966; Pelc and Welton, 1966; Woese et al., 1966; Crick, 1967, 1968; Ralph, 1968; Woese, 1968; Ohta and Kimura, 1971; Nagyvári and Fendler, 1974; Walker, 1974; Chernavskii and Chernavskaya, 1975; Hoffmann, 1975; Ishigami and Nagano, 1975; Lacey, Weber, and White, 1975; Mikelsaar, 1975; Ishigami, Nagano, and Tanatsuka, 1977; Nelsestuen, 1980; Reuben and Polk, 1980; Balasubramanian, 1982; Crothers, 1982; Jurka, Kolosza, and Roterman, 1982; Root- Bernstein, 1982; Grafstein, 1983; etc.). Their majority tries to solve the problem of the origin of biological information on the basis of a selected aspect (e.g., molecular structure, properties of amino acids, code relationships). However, there is not a single hypothesis among them, which would outline the subject together with a comprehensive prebiotic evolution theory or would incorporate it into such a theory.

A more or less comprehensive qualitative theory for prebiotic evolution has been presented by Hans Kuhn (Kuhn, 1972, 1976; Kuhn and Kuhn, 1978; Kuhn and Waser, 1981; Lehmann and Kuhn, 1984). The main difficulty of his theory lies in its macromolecule-centered thinking, thus he does not even try to outline a prebiotic metabolism; he accepts as a starting point that the monomers of nucleic acids and proteins have been provided on primitive Earth, and he investigates how from these monomers nucleic acids capable of evolution, then the genetic code, and through this, the protein-synthesizing apparatus could be formed. He realizes that other conditions are needed for the synthesis on a template than for the separation of the newly formed strand and the template; therefore he assumes that the possibility for this has been provided by periodic, program-like changes in the temperature (e.g., day and night, sunshine or shade). However, by this concept he turned the proliferation cycle into an externally controlled one, in contrast to the automatic internal control present in living systems and chemotons (see in Chapter IV).

According to Kuhn, the co-evolution of two types of macromolecules resulted in the formation of living systems. One (P_I) carries the genetic information from generation to generation, and the other (P_{II}) has a membrane-forming capacity; the latter forms the sack in which the former are placed. At increasing temperatures P_I molecules dissociate and serve as templates for the polymerization of monomeric P_I diffusing into the sack. At decreasing temperatures, however, molecules aggregate and the aggregates catalyze the formation of molecules from monomers. The molecules formed agglomerate to sacks around aggregates (Kuhn and Kuhn, 1978). Thus a functional relationship is established between the formation of P_I and P_{II} molecules. Though it is mentioned that the sack becomes unstable due its increased size, and that it stabilizes itself by fission, no explanation has been provided for this mechanism.

Further on, these macromolecules undergo evolution through mutations and selection. An outstanding role is played by the hairpin structures, as they can combine with single-stranded RNAs by triplets like tRNA, and thus they may be the primordial adapter molecules for the directed protein synthesis starting from amino acids (Kuhn and Waser. 1981). The authors did not realize that the probability for the formation of these hairpin structures is also exceptionally high (Gánti, 1979c).

Quantitative theoretical models

The first theoretical models with a claim to be quantitative, the hypercycle and chemoton models were born independently and both were published at the same time, in October 1971 (Eigen, 1971; Gánti, 1971). There are many similarities and numerous basic differences between these two models. Both models discuss functional organization, cycles and autocatalytic cycles play very important roles in both, both are based on the organization of chemical interrelations between autocatalytic cycles, and template processes have a basic role in both of them.

In contrast, in the hypercycle only functional relationships between macromolecules are considered, whereas in chemotons small molecules are also organized into a metabolic network; in the hypercycle model two different kinds of marcromolecules are supposed (information-carrying RNAs and catalytic proteins), whereas in chemotons there is only one (information-carrying) which, during evolution, develops to macromolecules having two different functions (information-carrying and catalytic), from which the second type of macromolecules specified to catalytic functions, develops only in the course of further evolution. In the hypercycle, functional relations only mean catalytic effects, in the chemoton these are basically stoichiometric (and kinetic) relations, while the catalytic relations develop only as a result of the evolution of chemotons.

In the hypercycle compartmentalization is not resolved, whereas in chemotons it is an organic part of organization and leads automatically to the separation of the new system formed by self-reproduction, i.e., to real proliferation. The formation of the hypercycle supposes the prior, spontaneous formation of nucleic acids having a very special monomer sequence, the formation of chemotons is possible with nucleic acids of arbitrary sequences, and the evolution of the chemoton results in the evolution and accumulation of nucleic acids with the specific information. Finally, the behavior of the chemoton model is studied in detail by the analysis of equation systems consisting of stoichiometric and kinetic equations of elementary chemical steps of participating reactions; whereas the "individual" behavior of the hypercycle has not been investigated to such detail (King's approach will be discussed later). As to the behavior of the hypercycle, Eigen and his co-workers assume that the kinetics of its growth is hyperbolic rather than exponential. Based on this assumption, they studied the evolutional behavior of "hypercycle populations" with a great mathematical apparatus. No similar investigations have been carried out for chemoton populations as yet.

The comparison of the hypercycle and chemoton theories is found in the work of Szathmáry (1984).

On designing the hypercycle theory, the main problem of Eigen concerned the origin and early evolution of biological information. He studied this question by examining what selection processes take place in "populations" of macromolecules capable of sequencetrue replication (i.e., of proliferation at the molecular level), as a consequence of differences in proliferation and decomposition processes that are due to faults in the replication. Assuming the system to be "quasilinear", by means of mathematical derivations of population dynamical character he concluded that such an "evolutional" process cannot lead to the formation of biological information. For this, a cooperative function is needed between the individual groups of macromolecules. Eigen examined four hypothetical cooperative models and found that only the hypercycle corresponded to evolutional conditions (Eigen, 1971).

Eigen's original representation of the hypercycle is shown in Fig. 6.2. For a better understanding, in Fig. 8.1 the representation of a hypercycle with four members is illustrated (based on Fig. 8 by Eigen and Schuster, 1979). Such a hypercycle contains nucleic acids with four nucleotide sequences (I_1 – I_4) and four protein enzymes with different sequences (E_1 – E_4). Circles with arrows around I_i indicate that the I_i molecules are not the direct elements of the hypercycle, but the autocatalytic cycles replicating the I_i molecules by template polymerization. Replication occurs enzymatically: the replication of I_i molecules is catalyzed E_{i-1} (I_1 in the case of E_4). In turn, the information on the sequence of E_i is carried by I_i.

Thus the operation of the hypercycle with four members is the following: RNA I_1 replicates with the aid of E_4 and then synthesizes enzyme E_1. This replicates I_2 according to the sequence by which enzyme E_2 is synthesized. This replicates I_3 and E_3, in turn replicates I_4 and E_4, and then the process starts again with a double amount of components.

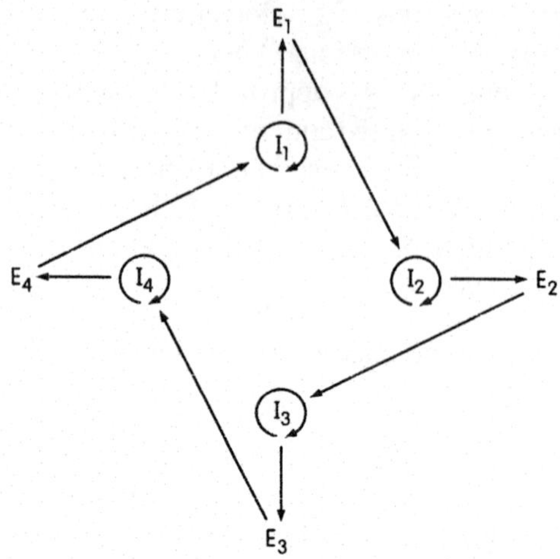

Fig. 8.1. A hypercycle with four members (after Eigen and Schuster, 1979).

According to Eigen and Schuster, the hypercycle is autocatalytic at two levels of organization, because the replication processes that are the elements of the system are autocatalytic by themselves, and on the other hand, as each element of the cycle of the hypercycle is a catalyst for the formation of the next element, the hypercycle itself is also autocatalytic, independently of its elements (Eigen and Schuster, 1977, 1978a, 1979). From this, they draw the conclusion that the type of the kinetics of the hypercycle should be

$$x=kx^2$$

i.e., it should be hyperbolic rather than exponential (Eigen and Schuster, 1978a, 1979). This hyperbolic character serves as the basis of calculation of evolutional processes.

The Eigen—Schuster theory based on the hypercycle model is a very fascinating intellectual construction. It points out that the evolutionary theory of Darwin can be applied also at the level of macromolecules, and that it can even be treated in an exact mathematical way. For the experimental checkup of the processes, the authors used the data obtained in the *in vitro* replication of the RNA of the Qß phage, and obtained very valuable results (Eigen and Schuster, 1979; Küppers, 1975, 1979). They analyzed the evolutional behavior of hypercycle populations by modern mathematical methods. Finally, they carried out numerous theoretical studies concerning the origin and properties of primitive genetic code (Eigen and Schuster, 1978b, 1979; Eigen et al., 1981; Eigen and Winkler-Oswatitsch, 1981a, b). This mathematical treatment was successfully developed further by Demetrius (1983) for the study of the genealogy of polynucleotides, by investigating also the effects of environmental changes on the evolution of the "quasi-species" of polynucleotides. The hypercycle model and the theory built on it had a great impact on scientific thinking, because this was the first widespread theory pointing out that prebiotic processes can be treated quantitatively on the basis of exact models. It influenced not only the way of thinking of researchers dealing with biogenesis, but also that of numerous other biologists, physicists and chemists. It gained adherents and opponents (Bresch, Niesert, and Harnasch, 1980; King, 1981; Szathmáry, 1984) and catalyzed the birth of new models (White, 1980; White and Raab, 1982; Matsuno, 1981; King, 1982).

In spite of all this, basic problems have arisen concerning the hypercycle model. In addition to its shortcomings already mentioned, namely, that it is not capable of explaining either the problem of metabolism or that of compartmentalization, the main difficulties are the following:

1. Though the main idea in designing the hypercycle model is to provide an explanation for the origin of the information contained in the monomeric sequence of macromolecules, it assumes the simultaneous and spontaneous formation of macromolecules containing information specialized to such an extent that is not known in biology even at present. A sequence dependency of such specificity and to such an extent as supposed for the hypercycle is not known among RNA polymerases.

2. In the living world there is no example for the hypercycle organization. In the only example cited by the authors, i.e., the "hypercycle with one member," the intercellular "proliferation" of the Qß phage is not a hypercycle on the one hand, as it does not possess the most characteristic feature of a hypercycle (the enzyme coded by the information carrier does not catalyze the formation of the next member of the chain), and on the other hand, the "proliferation" of the phage cannot be regarded as real proliferation, as has been shown in Chapter IV.

3. Not only is the model incapable of the origins of biological compartmentalization, but the hypercycle itself cannot operate without biological compartmentalization (Eigen et al., 1980; Bresch et al., 1980). Thus the model in its present form cannot be considered a complete model of an operative system.

4. The proliferation of the hypercycle is not hyperbolic. This error arises because the authors assume autocatalysis at two levels of organization, though at the second level in the hypercycle not an autocatalytic cycle but a normal cyclic process is operating which ensures the exponential proliferation of the elements of the first level. The proliferation of I_i-s is autocatalytic only E_{i-1} polymerizing them are present in excess, or if the proliferation of I_i-s and E_{i-1}-s run in parallel. The cyclic feedback at the second level of the hypercycle ensures just this latter condition, i.e., the exponential proliferation of the system.

If no concentration increase of E_{i-1} were ensured by this second feedback, the replication of I_i would not be of an exponential nature, because due to the continuous decrease in the ratio I_i: E_{i-1} the time needed for the replication of I_i molecules would increase also continuously. Thus under optimum conditions, the hypercycle shows exponential kinetic behavior. This means, at the same time, that the population-dynamic studies based on hyperbolic proliferation are not valid for the hypercycles.

Metabolic Network of a Prebiotic Chemoton

General aspects

In what follows, a metabolic network consisting of concrete prebiotic reactions will be designed from available experimental data, which in every respect corresponds to the abstract network of the chemoton model, and which is complete in the sense that it produces each intermediate, each internal component of the network by itself, only with a few compounds serving as nutrients taken up from outside. About two-thirds of the reactions involved are experimentally proved prebiotic reactions, and one-third of them are assumed but chemically possible reaction steps. It should be emphasized that this network is not the only possible network of chemoton coupling consisting of prebiotic reactions; at present one cannot even say that this is the most probable one. It should also be noted that due to the gaps in our knowledge concerning prebiotic reactions, neither may it be stated that all the reaction steps shown proceeded under prebiotic conditions and that they in fact have taken place.

The aim of designing and showing this network is to point out that the abstract chemoton model is a real model of prebiotic living systems and by introducing it, the process of the genesis of life can be understood and discussed in detail quantitatively. Nevertheless, it may be assumed that the prebiotic chemoton network shown here corresponds in its main features to the properties of the first living systems, even if not in all its details.

This network should contain concrete chemical autocatalytic processes corresponding to the metabolic, information, and membrane subsystems, together with linear synthesis pathways coupling these subsystems stoichiometrically, as well as other chemical processes ensuring the operation of the whole system (Fig. 8.2).

In selecting nutrients, two basic aspects should be considered: nutrients should be very frequent compounds under prebiotic conditions and they should diffuse through the membranes of prebiotic chemotons without an active transport mechanism.

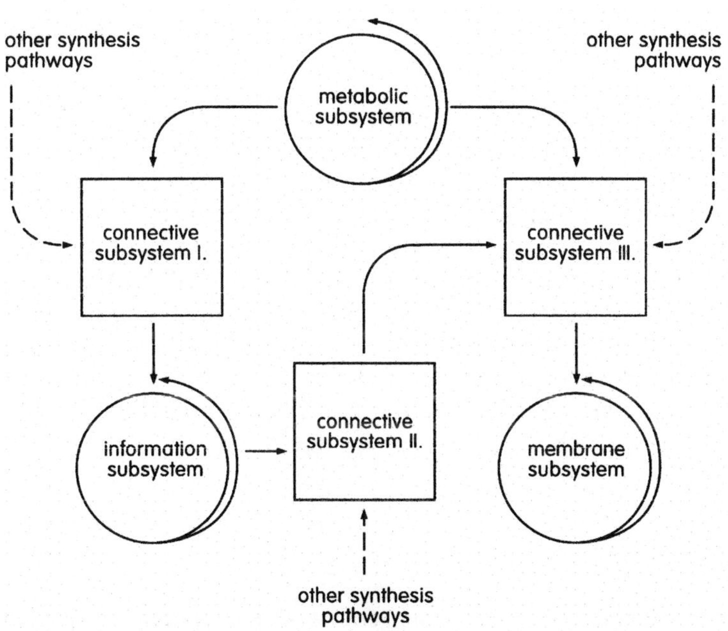

Fig. 8.2. When constructing the metabolic network of a prebiotic chemoton, concrete prebiotic chemical systems corresponding to the subsystems shown should be looked for and they should be connected through appropriate stoichiometric couplings.

The first requirement is easy to satisfy: chemical evolutional experiments prove the outstanding importance of formaldehyde, cyan and cyano derivatives, and ammonia. At the same time, radio-astronomical studies also find these compounds to be the most frequent C and N compounds in the Universe, and according to the results of chemical evolutional experiments, the same substances are formed from other simple C- and N-containing compounds as early intermediates when appropriate amounts of energy are supplied.

These compounds also satisfy the second requirement. Namely, according to the chemoton model, the surrounding membrane is a two-dimensional liquid with a double molecular layer consisting of amphipathic molecules the interior of which contains hydrophobic chains, but it is covered by electrically charged groups on both of its sides. Water-soluble molecules can penetrate this membrane only through momentarily formed holes between the membrane-forming molecules caused by thermal motion. Therefore small such molecules as HCHO, HCN, NH_3, and H_2O can diffuse through the membrane easily, but products formed from them, i.e., the intermediates of the metabolism of the chemoton network, only with difficulty or not at all. Thus the precondition of the operation of chemotons – that the membrane should be permeable for nutrients (and waste materials), but must not be permeable for network intermediates or internal substances – may be fulfilled without specific transport mechanisms.

This effect may be even stronger if the intermediates are electrically charged, because their interaction with the charge of the membrane may further hinder their diffusion through the membrane. For example, Moiseeva studied the diffusion rates of sugars and sugar phosphates through polymeric membranes and found that the diffusion of sugar phosphates is by 1–2 orders of magnitude smaller than that of the corresponding sugars (Moiseeva, 1971). It cannot be accidental that the intermediates of the metabolic network of living beings are almost without exception ionic compounds, and that e.g., the first step of the utilization of sugars is phosphorylation, and in an extraordinarily fast, practically irreversible reaction at that. This leads to the immediate formation of charged sugar phosphate ions for which the membrane is impermeable.

Autocatalytic subsystems

For the selection of metabolic subsystems, chemical evolutional experiments carried out so far would suggest that the intermediates of the metabolic network should be amino acids and organic acids, because on the basis of these experiments, their spontaneous formation seems to be the easiest. In addition, amino acids are indispensable constituents of proteins and organic acids are capable of forming a self-reproductive chemical cycle, i.e., the reductive carboxylic acid cycle, which could ensure the self-reproducing nature of the metabolic subsystem.

However, the reductive carboxylic acid cycle is only present in the metabolism of a small fraction of living beings. It can be rightly assumed that the most general part of the metabolic network present in all living beings originates from the most primitive network. In the metabolism of living systems a major part of protein-forming amino acids is synthesized not in the network, but in coupled linear synthetic pathways (cyclic amino acids, threonine, lysine, etc.) which means that these synthetic pathways have not been parts of the primitive metabolic network, but they have been developed only later, by evolution. This assumption is supported by the fact that the formation of mainly these amino acids could not be detected in the general chemical evolutional experiments. It is hardly probable that the complex genetic code and translation apparatus were formed earlier than the possibilities for the synthesis of all of the 20 protein-forming amino acids. In contrast, the central part of the metabolism of every living being known is the carbohydrate metabolism, for both anaerobic and aerobic metabolism. The formation of amino acids is also based on sugar metabolism; the citrate cycle operates by utilizing the acetyl-CoA formed in sugar metabolism, the photosynthesis of plants builds CO_2 into organic compounds by means of the self-reproducing Calvin cycle consisting of the transformation network of sugar phosphates, thus providing the biochemical basis for the conditions of growth in the present world of living systems.

Thus the study of the metabolic network of the present world of living beings suggests the priority of carbohydrate metabolism rather than that of amino acids and proteins. This recognition provides us with a key to the solution: the formose reaction can be chosen as the basic part of the metabolic network of primitive chemotons (Gánti, 1978c, 1979c, 1980). This choice is very advantageous, because:

1. Formaldehyde is one of the most frequent prebiotic substances in the Universe and it was present in the primitive atmosphere (*see* Chapter VII).

2. The formose reaction represents an autocatalytic, self-reproducing network (Cairn-Smith and Walker, 1974; Decker, 1974; Mizuno and Weiss, 1974; etc.).

3. Both experimental results and theoretical considerations showed that the formose reaction could take place under prebiotic conditions (Reid and Orgel, 1967; Gabel and Ponnamperuma, 1967; Decker and Speidel, 1972; Decker, 1974, 1978; Cairn-Smith and Walker, 1974; etc.).

4. Through glyceraldehyde formed in this reaction a stoichiometric connection may be established to the synthesis of membrane-forming molecules (Gánti, 1978c, 1979c).

5. Through ribose also formed, the synthesis of the raw materials of template polymers may be realized (Gánti, 1978c, 1979c).

6. Via the intermediates of the reaction, a stoichiometric possibility may be provided for the formation of branched isoprenoids, as well as of valine and leucine (Decker and Schveer, 1983).

7. Formaldehyde is indispensable in most of the experiments for the formation of microspheres (Bahadur, 1964, 1966, 1967; Bahadur and Ranganayaki, 1983; Fraser and Folsome, 1975; Folsome, 1976, 1977; Folsome and Brittain, 1981; Kenyon and Nissenbaum, 1976; etc.).

8. By the formose reaction, every sugar playing a role in metabolism could be formed in spontaneous reactions without enzymes (Cairn-Smith and Walker, 1974; Mizuno and Weiss, 1974).

9. Formaldehyde could be formed in the primitive atmosphere continuously as a result of the UV radiation of the Sun; thus the inverse assimilation cycle of Decker could develop, which provided continuously for the nutrient supply of the proliferating chemotons (Decker, 1974, 1978).

For all of these reasons, the formose reaction is chosen as the metabolic subsystem of the prebiotic chemoton model. The concrete network is shown in Fig. 8.3. It should be noted that in the figure no alternative reaction pathways are shown for the sake of better understanding.

In selecting the information subsystem, there are not many chemical processes to choose from, because up to now only four different template replication processes are known: the template polycondensations of DNA and RNA, the template replication processes of clay minerals (Weiss, 1981), and the template polymerization of acenaphthylene (Cser, Hardy, and Veksli, 1974).

The replication of clay mineral templates cannot be coupled chemically in organic metabolism, thus it cannot here play the role of the information subsystem. The mechanism of the template polymerization of acenaphthylene is completely unknown. Since the template in this case is built of only one kind of building unit (acenaphthylene), its capacity as information carrier would be very small, and therefore would not be a suitable choice.

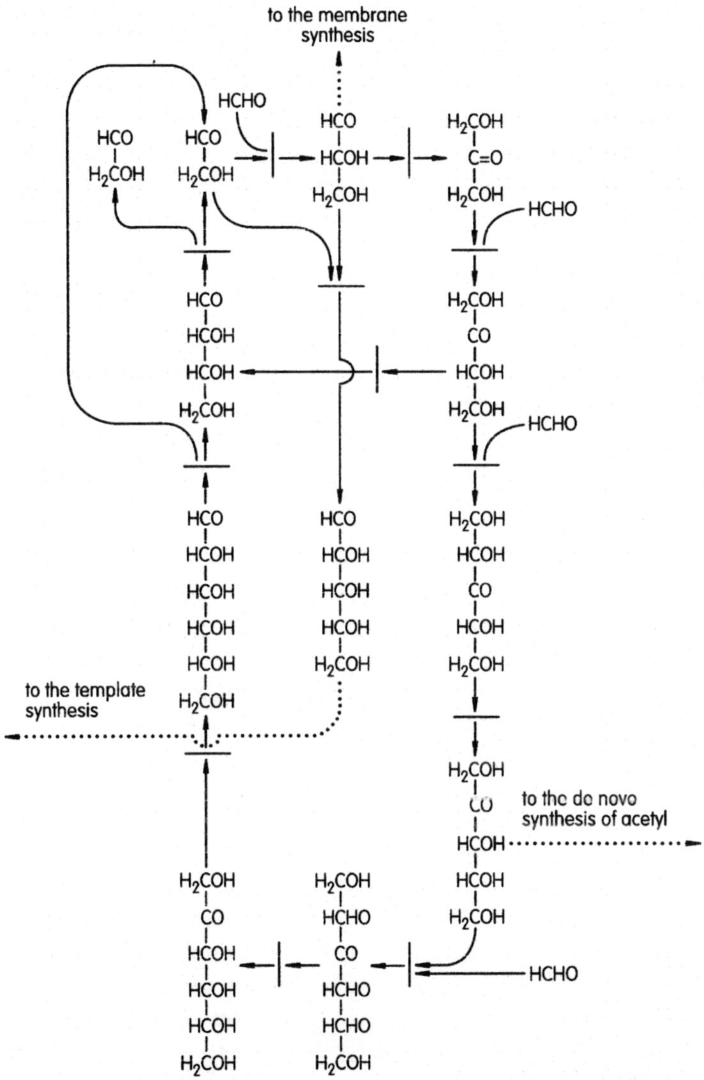

Fig. 8.3.The formose network chosen as the metabolic subsystem of the concrete prebiotic chemoton model. The selected sites for the coupling to the other subsystems are also shown.

As DNA is the general genetic information carrier in the living world, it seems to be only too natural to choose DNA as the information subsystem of prebiotic chemotons. However, neither deoxyribose nor thymine appears among the chemical evolution products [though it was shown that from uracil, 5-hydroxymethyl uracil could be formed with formaldehyde, which could be reduced to thymine with formic acid (Chadha and Choughuley, 1983)]. Thus the direct prebiotic formation of DNA is not very probable. The too rigid structure of DNA also speaks against it, as this makes the direct utilization of the information content carried by its sequence impossible without a translation mechanism. As we will discuss later, this is possible in a direct way in the case of RNA.

Thus it is obviously expedient to choose the RNA template polycondensation process as the information subsystem of prebiotic chemotons. It is a generally accepted view among researchers concerned with biogenesis that the appearance of RNA preceded that of DNA in the course of evolution. Ribose, and purine and pyrimidine bases needed for the synthesis of RNA are among the reaction products of chemical evolution as we have shown in detail in the previous chapter. Stoichiometric coupling with the formose cycle may be ensured by ribose.

However, the mechanism of the template process could not have been identical with that of the present. First, as already pointed out, a systematic formation of nucleoside triphosphates cannot be assumed under prebiotic conditions. Second, the organic chemical synthesis of oligonucleotides cannot be realized in the same way as the enzymatic synthesis. As condensing agents, mainly dicyclohexyl carbodiimide was used by Khorana and Todd (Khorana and Todd, 1953; Khorana, 1961); others employed acid chlorides, trichloroacetonitrile, Woodward's reagent, etc. A significant step forward was Cramer's discovery, i.e., the use of nucleoside imidazoles in the synthesis of oligonucleotides (Cramer, 1961, 1966). Nowadays the reproduction experiments of prebiotic RNA synthesis are either carried out in the presence of imidazole (e.g., Ybanez et al., 1971a; Lohrmann and Orgel, 1978), or the starting materials are nucleoside imidazolides (e.g., Sleeper, Lohrmann, and Orgel, 1978; Sleeper et al., 1979; Sleeper and Orgel, 1979). Imidazole, as was shown in the previous chapter, is also formed in prebiotic reactions. Based on these, findings the template reaction of ribonucleotide imidazolides seems to be the most convenient choice for the information subsystem of prebiotic chemotons.

Most uncertainties are involved in the selection of the membrane subsystem, as experimental data concerning the prebiotic formation of lipids are most scarce (see Chapter VII). At present, the possibility cannot be excluded either that the material of primitive compartments consisted of phospholipid membranes, similarly to the membranes of eubacteria and eukaryotic cells (Deamer and Oró, 1980).

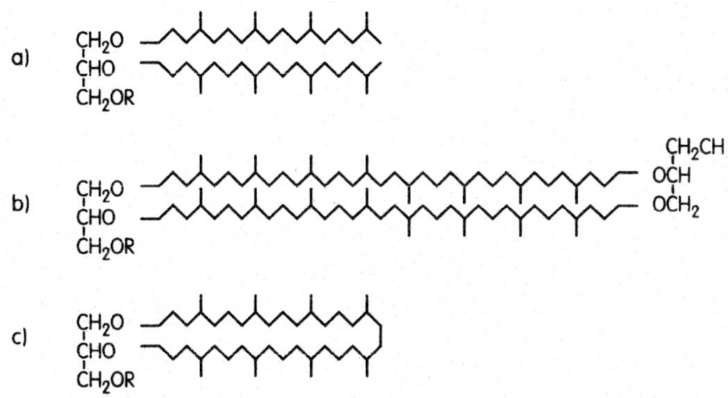

Fig. 8.4. The building stones of the membrane of archaeobacteria are not phospholipids but the alkylethers of glycerol: a) bis(phythanil) glycerol diether; b) bis(dyphythanyl) diglycerol tetraether; c) macrocyclic alkyl glycerol ether.

However, the existence of archaebacteria warns us to be careful. The cell membrane of these microorganisms, considered presently to be the most ancient ones, consists of alkyl ethers of glycerol rather than of glycerol esters of fatty acids. The studies of the membrane of methanogens (Makula and Singer, 1978; Tornabene et al., 1978; Tornabene and Langworthy, 1979; Kushwaha et al., 1981; Langworthy, Tornabene and Holzer, 1982; Comita and Gagosian, 1983), as well as that of other archaebacteria (DeRosa et al., 1980; Ross et al., 1981; Langworthy et al., 1982) have proved this. In archaebacteria the most frequent building units of membranes are: (1) bis(phythanyl) glycerol diether; (2) bis(diphythanyl) diglycerol tetraether; and (3) in the case of methanogen bacteria living in deep-sea thermal well (Comita and Gagosian, 1983), a macrocyclic alkyl glycerol ether (Fig. 8.4). Thus the suggestion of Hidvégi (1980) to choose glycerol diphythanol ether or more precisely, its imidazole-derivative as the basis of the membrane subsystem for the prebiotic chemoton, seems to be acceptable (Fig. 8.5).

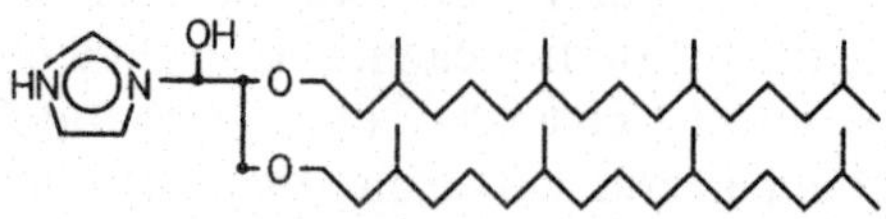

Fig 8.5. The compound formed from glycerol diphythanyl ether and imidazole has been chosen as the material of the membrane of prebiotic chemotons as suggested by Hidvégi.

Connective subsystems

Theoretically, three connective reaction pathways are needed for connecting the metabolic subsystem to the information and membrane subsystems and coupling the latter two to each other. In practice, the reaction pathways coupling the information and metabolic subsystems vary depending on whether the synthesis of purine or pyrimidine nucleotide is required. In what follows, the reaction pathways proposed by Hidvégi will be discussed.

Connective subsystem I/A starts from ribose formed in the formose cycle; this is phosphorylated in a Halmann's reaction, then upon its reaction with a purine base, a purine nucleotide is formed. A subsequent reaction with cyanamide and coupling with another nucleotide result in a NAD-like dimer. The P—N bond is formed in the next step via the reaction between the dimer and formamide. One nucleotide is liberated (thus it plays a catalytic role in the reaction) whereas on the N atom connected to the other nucleotide, imidazole ring formation takes place by the addition of two molecules of formaldehyde and one ammonia (Fig. 8.6).

Connective subsystem I/B is a hypothetic pathway of the abiotic synthesis of pyrimidine nucleotide imidazolides. The second part of the pathway, the formation of imidazolide, is identical with the former one, but the formation of the nucleotide is completely different, which also depends on whether cytidyl or uridyl nucleotide is formed. The hypothetic pathways are illustrated in Figs. 8.7 and 8.8.

Connective subsystem II couples formose reaction and membrane formation. The glyceraldehyde formed in the formose reaction is phosphorylated by acetylphosphate, then it produces an ether bond with two geranyl-geranyl pyrophosphate molecules, and finally the geranol-geranol side chains are hydrogenated to phythanol side chains (Fig. 8.9).

Connective subsystem III consists of a single reaction step in which the diphythanol ether synthesized in connective subsystem II reacts with imidazole liberated in template polymerization (Fig. 8.10).

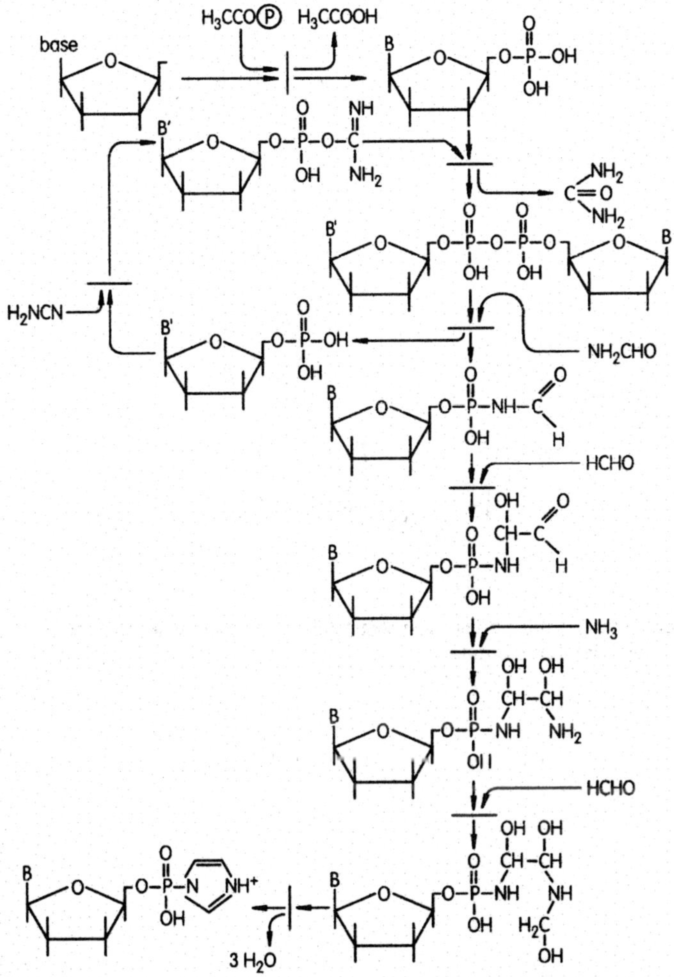

Fig. 8.6. One of the reaction chains connecting the formose cycle and the synthesis of RNA is the supposed prebiotic synthesis of purine nucleotide imidazolides according to Hidvégi.

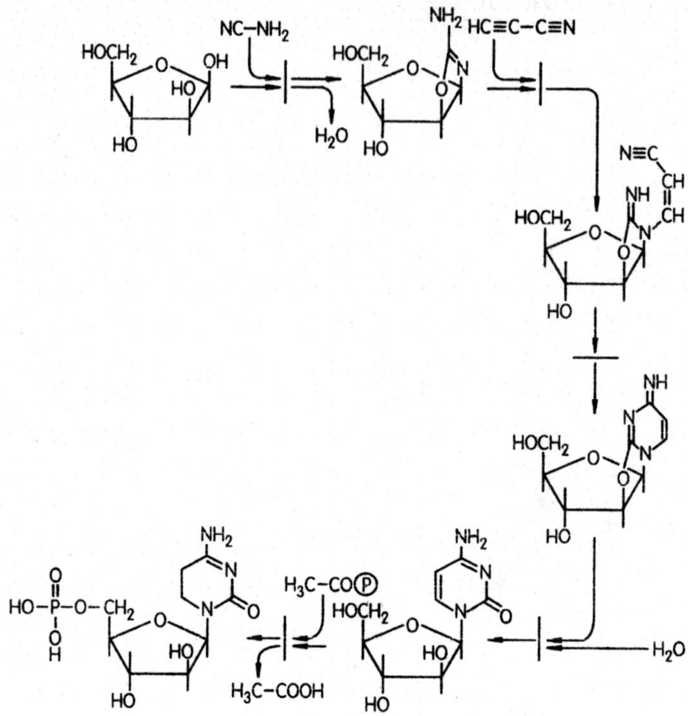

Fig. 8.7.The other hypothetic reaction chain connecting the formose cycle and the synthesis of RNA is the synthesis of cytidyl nucleotide.

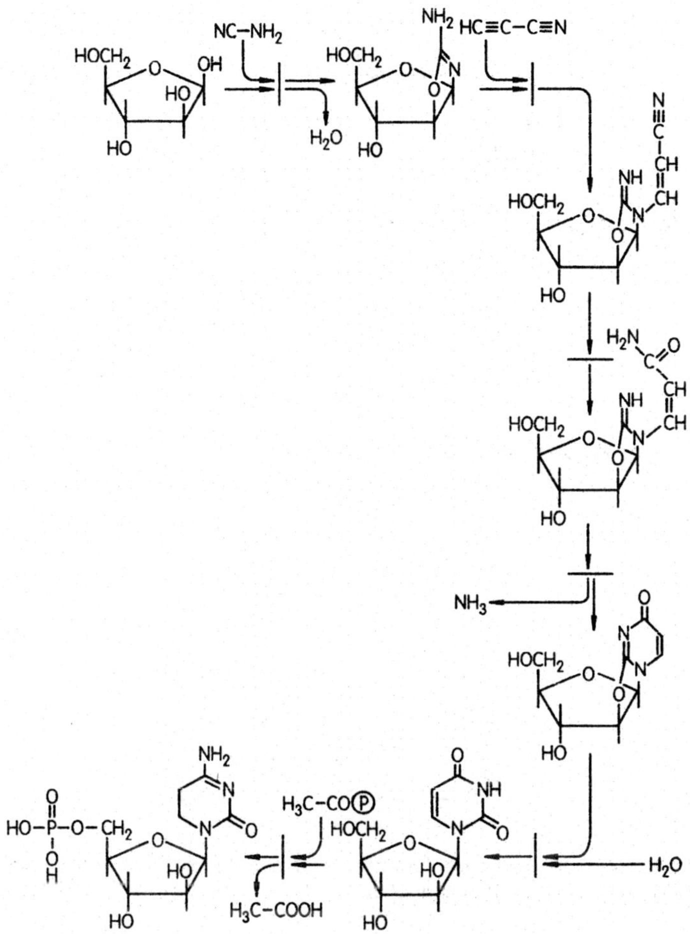

Fig. 8.8.The third hypothetic reaction connecting the formose cycle and the synthesis of RNA is the prebiotic uridyl nucleotide synthesis.

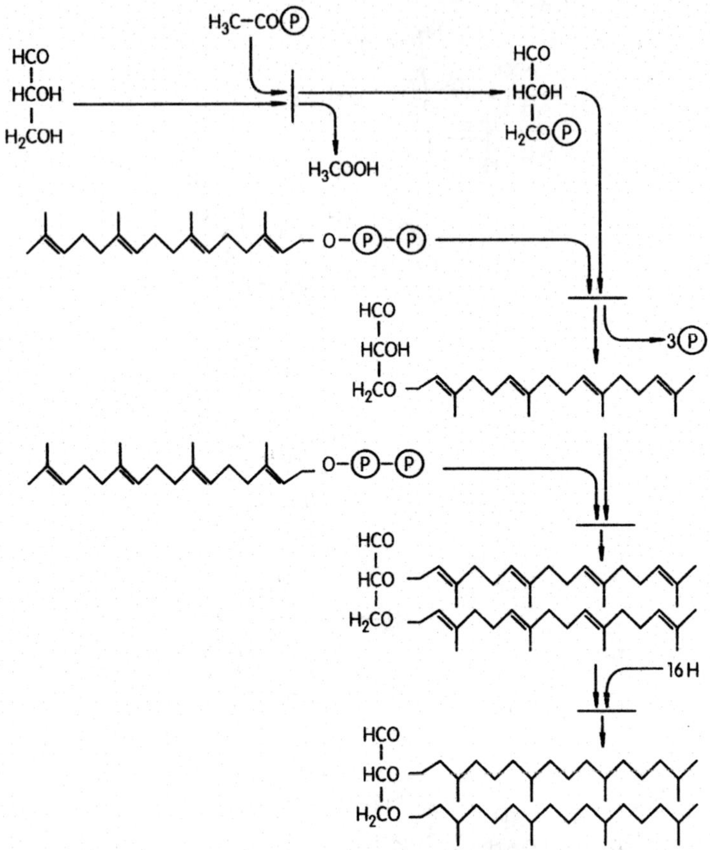

Fig. 8.9. Part of the pathway connecting the formose cycle and the membrane formation is, according to Hidvégi, the formation of diphythanol ether.

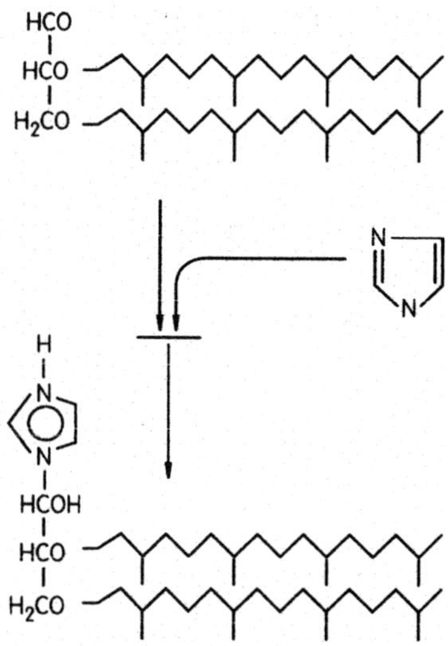

Fig. 8.10. The reaction of diphythanol ether and the imidazole liberated in template polymerization results in the building material of the membrane.

Additional pathways

By assigning the autocatalytic subsystems and the connective subsystems, the basic network is given for the metabolism of prebiotic chemotons. This system could satisfy already the chemoton function if all of the substances were also provided that would be used as nutrients in the different reactions. However, it is obvious that among these compounds there are substances that could not be utilized as nutrients, because the membrane is not permeable for them. Such are the purine bases, geranyl-geranyl pyrophosphate, and acetylphosphate. The system has to synthesize these substances from real nutrients internally. This task is performed by the additional reaction pathways.

The synthesis of purine bases can take place via reaction pathways shown in Figs. 7.3 and 7.4.

There are no experimental data concerning the prebiotic synthesis of geranyl-geranyl pyrophosphate. In order to complete the network, Hidvégi simply assumed that the synthesis takes place under prebiotic conditions analogously with the biological synthesis if acetylphosphate is applied as both carbon source and phosphorylating agent. This seems to be acceptable unless contradictory experimental data become available. The hypothetic reaction pathway thus obtained is shown in Figs. 8.11 and 8.12.

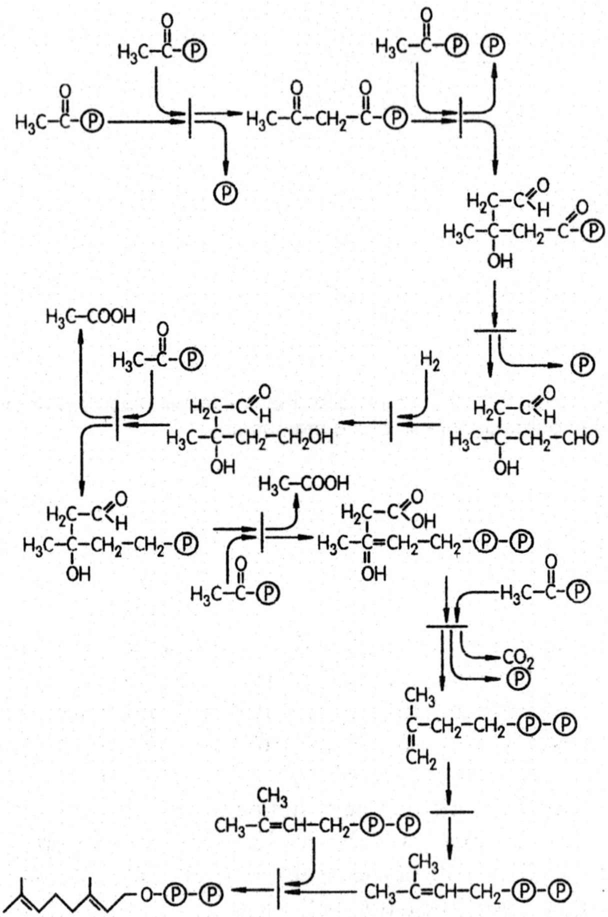

Fig. 8.11. The prebiotic formation of geranyl pyrophosphate as suggested by Hidvégi.

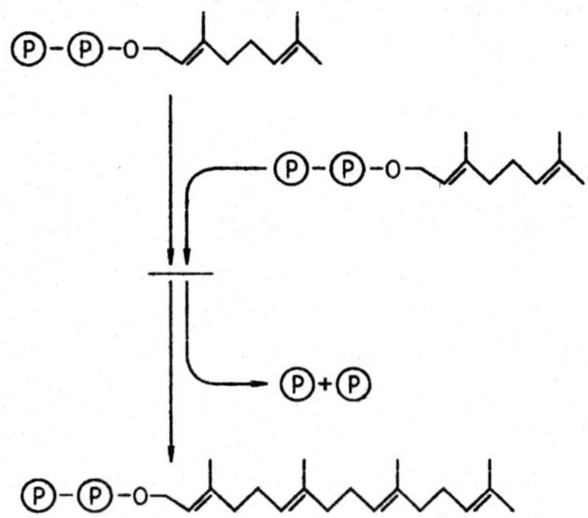

Fig 8.12. The formation of geranyl-geranyl pyrophosphate.

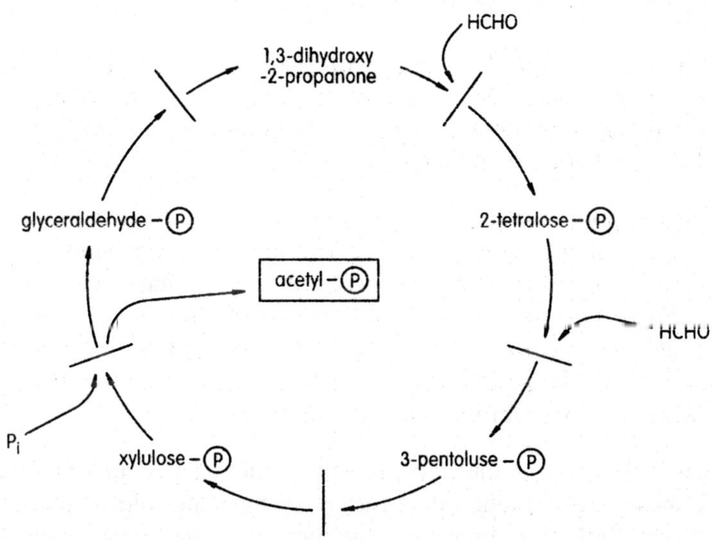

Fig 8.13. The prebiotic cycle of the *de novo* synthesis of acetylphosphate as suggested by Hidvégi.

Acetylphosphate plays two roles in the prebiotic chemoton network: it serves as both acetylating and phosphate donating agent. To clarify the acetylating function in the *de novo* synthesis of acetylphosphate, we may start from a biological example: in prokaryotes phosphoketolase is accepted as the enzyme catalyzing the reaction

$$\text{xylulose-5-P+P}_i \rightarrow \text{acetyl-P+glyceraldehyde-P}$$

The unusual mode of the utilization of inorganic phosphate and the occurrence of this enzyme in prokaryotes suggest that this reaction could perhaps be a reaction of primitive metabolism. Considering that the active center of this enzyme contains an imidazole group, it is possible that the imidazoles bound to the membrane of prebiotic chemotons could also have catalyzed this reaction. Because xylulose is formed in the formose reaction from glyceraldehyde and one of the products is also glyceraldehyde, the formation of an "acetyl-P-producing cycle" seems probable. Thus the acetylphosphate synthesis becomes stoichiometrically independent of the formose cycle (Fig. 8.13).

A qualitative survey of the operation of the network

In the possession of concrete (partly hypothetic) reaction systems for all three autocatalytic subsystems and the connective pathways satisfying the requirements of the abstract chemoton model, let us outline first the construction and operation of the hypothetic prebiotic chemoton.

We take the size of the prebiotic chemoton spheres to be identical with that of other microspheres, i.e., about 1 μm diameter. The sphere is surrounded by a two-dimensional membrane consisting of glycerol ethers that also contain imidazole groups. The membrane is permeable for small molecules and ions (consisting of 2–5 atoms); medium-size molecules (consisting of 5–50 atoms) can diffuse through the membrane but ions of the same size cannot, whereas for large molecules and ions the membrane is equally impermeable.

All of the raw materials used in the network (formaldehyde, ammonia, cyan, phosphate) satisfy these requirements, thus they may play the role of nutrients for the chemoton. At the same time, for the products participating in the construction of the chemoton, the membrane is impermeable, and thus they may serve as internal components. Imidazole combining with the membrane provides space for speculations: it cannot be excluded that with its assistance a primitive active transport may also be realized (Stillwell, 1980).

Because of the small size of the spheres, the diffusion pathway of nutrients is very short, thus the nutrients used up in the reactions can be supplemented quickly. As the membrane is impermeable to the products of these reactions, the concentration of organic substances may increase independently of the concentration of organic materials in the extrachemotonial space. (However, it depends on the osmotic pressure which is regulated by the mechanism of division.) Thus prebiotic chemotons are capable of functioning also at very low nutrient concentrations!

Formaldehyde diffused into the chemoton is transformed to sugars (sugar phosphates?) in the formose reaction providing thereby the necessary raw materials for template reaction and membrane synthesis, ribose and glyceraldehyde. Ribose is transformed into ribonucleotide imidazoles via connective pathway I and the coupled additional reaction pathways. From glyceraldehyde, glyceraldehyde diphythanol ethers are formed via connective pathway II and the additional pathways, which are assumed to provide aggregates in the "chemoton plasm" as a result of their decreased stability in water.

In the course of this process, the concentration of ribonucleotide imidazoles keeps increasing in the internal space. When their concentration reaches a critical value (for more detail see Chapter IV), the double-helix structure of RNA molecules unfolds and the synthesis of a new strand begins on both RNA strands. In the reaction imidazole is liberated, which then reacts with the glyceraldehyde diphythanol ether molecules of the aggregates, thereby dissolving them. Subsequently, these membrane-building units spontaneously incorporate into the membrane thereby increasing the membrane surface. This increase in the surface area changes the concentration relations, and according to the mechanism of division described earlier, the prebiotic chemoton divides into two identical prebiotic chemotons, which are also identical to the original one, and the process starts again in both newly formed chemotons.

Quantitative Description of Prebiotic Chemotons

Strategy

Cyclic stoichiometry and stoichiokinetics make possible the detailed and complete quantitative description of the metabolism of the prebiotic chemoton previously outlined as well as the characterization of the given chemoton by a single overall equation. This overall equation tells we what subsystems and cycles (closed networks) are involved in the given chemoton, it provides the amount of intermediates, the quality and quantity of raw materials needed for growth and proliferation, and it shows the turning number of cycles and the quality and quantity of waste materials. The calculations for the prebiotic chemoton network presented in the previous chapter have been carried out by Szathmáry (1982). The scheme of the complete system is illustrated in Fig. 8.14.

The following strategy is applied in deriving the overall equation:

1. The overall equation for the synthetic pathway of the membrane-forming molecule is derived, considering the necessary amount of this molecule for doubling the membrane surface.

2. The overall equations for the synthetic pathways providing the nucleotide imidazolides needed for the replication of template molecules are derived.

3. The overall equations obtained in phases 1 and 2 are stoichiometrically coupled.

4. The equations for the pathways synthesizing the internal precursors are elaborated and the overall equations obtained so far are summed up.

5. The overall equation of the formose cycle is determined by considering the amounts of materials needed for the duplication of the cycle and for the synthesis pathways discussed earlier. The combination of this and the previous overall equations represents the overall equation of the prebiotic chemoton.

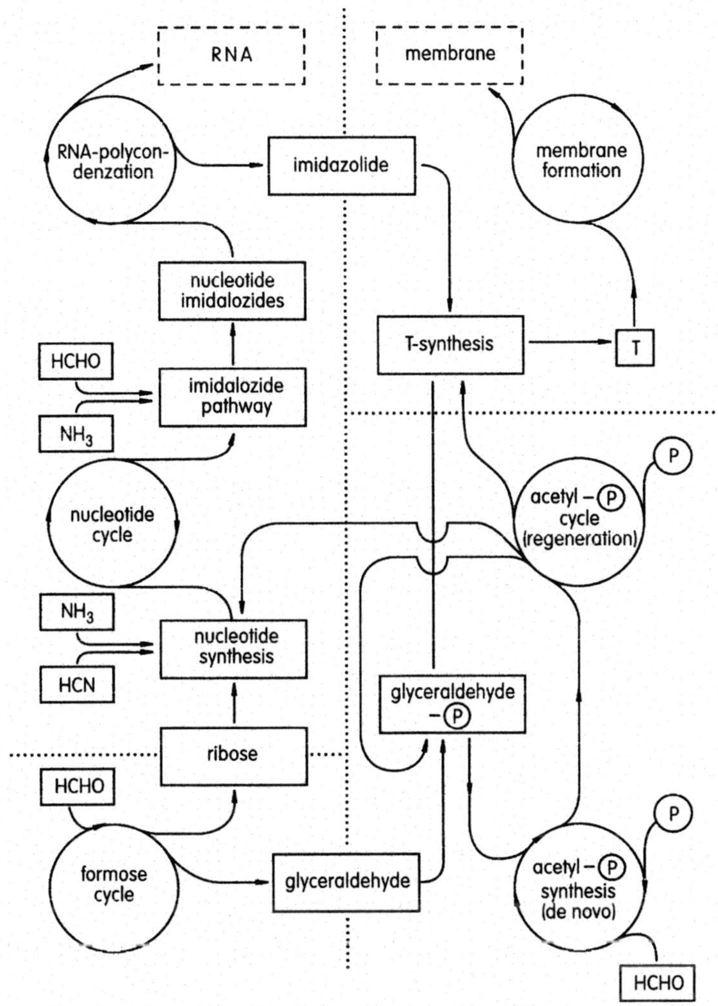

Fig. 8.14. The scheme of the complete metabolic network of the prebiotic chemoton.

Abbreviations used:

A	adenosine
AcAcP	acetoacetylphosphate
AcP	acetylphosphate
Ade	adenine
AHex	aldohexose

AICA	4-amino-imidazole-5-carboxamide
AICAI	4-amino-imidazole-5-carboxamidine
AMADAM	amino-malonodiamidine
AMANIL	amino-malononitrile
ATet	aldotetrose
C	cytidine
C^1–C^3	intermediates of cytidine synthesis
Cyt	cytosine
DiMetAlL-PP	dimethyl-allyl-pyrophosphate
FMADN	formamidine
G	guanosine
GADP	glyceraldehyde-phosphate
GGYL-PP	geranyl-geranyl-pyrophosphate
GlAD	glycolaldehyde
Gua	guanine
GYL-PP	geranylpyrophosphate
H^*	reducing agent
H_i	intermediate of ribose synthesis
Hu	hexulose
IAcNIL	imino-acetonitrile
Int	intermediate of membrane synthesis
IPYL-PP	isopentenyl-pyrophosphate
MEVAC	mevalonic acid
MEVAld	mevaldehyde
NS	nucleoside
OH-Met-GYL-P	ß-OH-ß-methyl-glutaryl-phosphate
$(OH)_2$–PON	1,3-dihydroxy-2-propanone
P_i	inorganic phosphate
-P	bound organic phosphate
-PP	bound organic pyrophosphate
PE	pentulose
R	imidazole
Rib	ribose
T	glycerol diphythanol ether imidazolide
Tu	tetrulose
U	uridine
U^3–U^4	intermediates of uridine synthesis
Ura	uracyl
Xu	xylulose

Membrane synthesis

The scheme of the synthesis of membranogen molecules (T) is outlined in Fig. 8.15. As the synthesis pathway does not contain any cyclic part, the overall equation may be deduced by the application of usual chemical stoichiometry.

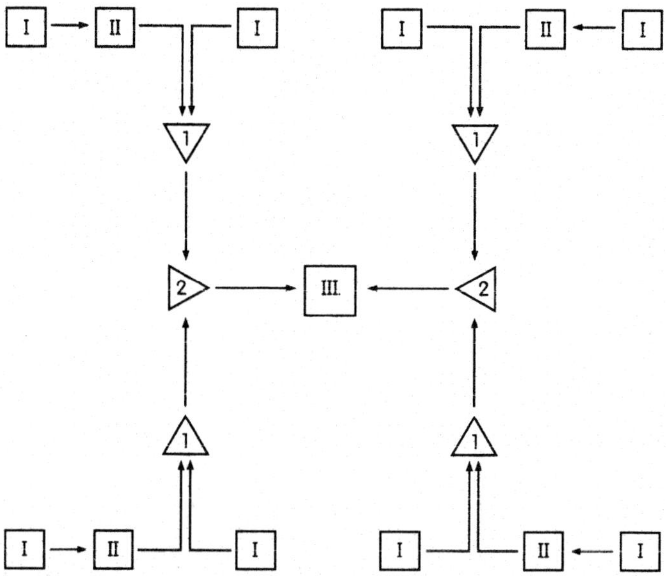

Fig. 8.15. The calculation scheme of the synthesis of membrane-forming molecules.

Branch I

2AcP	$\longrightarrow$	AcAcP + P_i	[8.1.1]
AcAcP + AcP	$\longrightarrow$	OH-Met-GYL-P + P_i	[8.1.2]
OH-Met-GYL-P + H^*	$\longrightarrow$	MEVAld + P_i	[8.1.3]
MEVAld + H^*	$\longrightarrow$	MEVAC	[8.1.4]
MEVAC + AcP	$\longrightarrow$	5-P-MEVAC + Ac	[8.1.5]
5-P-MEVAC + AcP	$\longrightarrow$	5-PP-MEVAC + Ac	[8.1.6]
5-PP-MEVAC + AcP	$\longrightarrow$	3-P-5-PP-MEVAC + Ac	[8.1.7]
3-P-5-PP-MEVAC	$\longrightarrow$	IPYL-PP + CO_2 + Pi	[8.1.8]
6AcP + $2H^*$	$\longrightarrow$	IPYL-PP + $4P_i$ + 3Ac + CO_2	[8.1]

H* denotes a reducing agent in general terms.

Branch II

$$\text{IPYL-PP} \rightleftarrows \text{DiMetAIL-PP} \qquad [8.2]$$

Branch III

$$\text{GADP} + \text{R} \rightleftarrows \text{INT} \qquad [8.3.1]$$

$$\text{Int} + \text{GGYL-PP} \longrightarrow \text{„Monoether”} + 3P_i \qquad [8.3.2]$$

$$\text{„Monoether”} + \text{GGYL-PP} \longrightarrow \text{„Diether”} + 2P_i \qquad [8.3.3]$$

$$\text{„Diether”} + 32H^* \longrightarrow \text{T} \qquad [8.3.4]$$

$$\text{GADP} + 2\text{GGYL-PP} + \text{R} + 32H^* \longrightarrow \text{T} + 5P_i \qquad [8.3]$$

The connection points of branches (denoted “1” and “2” in Fig. 8.15):

$$\text{DiMetAIL-PP} + \text{IPYL-PP} \longrightarrow \text{GYL-PP} + 2P_i \qquad [\,8.4\,]$$

$$2\text{GYL-PP} \longrightarrow \text{GGYL-PP} + 2P_i \qquad [\,8.5\,]$$

In order to obtain the overall equation for the synthesis of T molecules, all of the synthetic pathways in Fig. 8.15 should be summed up. Accordingly, multiplications should be carried out: for Eq. [8.1] by 8, for Eqs. [8.2] and [8.4] by 4, and for Eq. [8.5] by 2. The equations thus obtained plus Eq. [8.3] should be added together. This yields

$$48\text{AcP} + \text{GADP} + \text{R} + 48H^* \longrightarrow \text{T} + 8CO_2 + 49P_i + 24\text{Ac} \qquad [8.6]$$

If the membrane of prebiotic chemotons is built of m pieces of T molecules, the growth of the membrane from division to division is given by the cycle:

$$\boxed{T_m} + mT \xrightarrow{\;①\;} \boxed{T_{2m}} \qquad [8.7]$$

This should be added to Eq. [8.6] multiplied by m, and thus we obtain the stoichiometric overall equation for the membrane synthesis of prebiotic chemotons:

$$\boxed{T_m} + m(48AcP + GADP + R + 48H^*) \xrightarrow{①}$$

$$\boxed{T_{2m}} + m(49P_i + 8CO_2 + 24Ac) \qquad [8.8]$$

Template synthesis

The synthesis of guanine occurs via the following reaction steps:

$$2HCN \longrightarrow IAcNIL \qquad [8.9.1]$$

$$IAcNIL + HCN \longrightarrow AMANIL \qquad [8.9.2]$$

$$AMANIL + 2NH_3 \longrightarrow AMADAM \qquad [8.9.3]$$

$$HCN + NH_3 \longrightarrow FMADN \qquad [8.9.4]$$

$$AMADAM + FMADN \longrightarrow AICAI + 2NH_3 \qquad [8.9.5]$$

$$AICAI + H_2O \longrightarrow AICA + NH_3 \qquad [8.9.6]$$

$$AICA + (NC)_2 \longrightarrow Gua + HCN \qquad [8.9.7]$$

Because parts of NH_3 and HCN also appear among the products, they behave for the whole reaction sequence as catalysts; therefore cycle stoichiometry should be applied for the description of this reaction series. Thus the overall equation of linear pathway synthesis [8.9.1] – [8.9.7] will be a cyclic equation:

$$3HCN + (NC)_2 + H_2O + \underline{3NH_3} + \underline{HCN} \xrightarrow{①} Gua + \underline{3NH_3} + \underline{HCN} \qquad [8.9]$$

The first five steps of adenine synthesis are identical to those of guanine synthesis described by Eqs. [8.9.1] – [8.9.5]. The sixth step is:

$$\text{AICAI} + \text{FMADN} \longrightarrow \text{Ade} + 2\text{NH}_3 \qquad [8.10.1]$$

Considering that FMADN is formed in reaction [8.9.4], the overall equation of adenine synthesis is obtained upon the addition of Eqs. [8.9.1] – [8.9.5] and [8.10.1], to them [8.9.4] should be added once again. The result is:

$$5\text{HCN} + \underline{4\text{NH}_3} \xrightarrow{\text{①}} \text{Ade} + \underline{4\text{NH}_3} \qquad [8.10.2]$$

For the synthesis of purine nucleotide, ribose-1-phosphate is needed, which is synthesized from glycolaldehyde and phosphorylated in Halmann's reaction:

$$\text{GlAD} + \text{H}_2\text{CO} \rightleftharpoons \text{GAD} \qquad [8.11.1]$$

$$\text{GlAD} + \text{GAD} \rightleftharpoons \text{Rib} \qquad [8.11.2]$$

$$(\text{NC})_2 + \text{P}_i \longrightarrow \text{H}_i \qquad [8.11.3]$$

$$\text{H}_i + \text{Rib} \longrightarrow \text{Rib-1-P} + \text{NCCONH}_2 \qquad [8.11.4]$$

$$2\text{GlAD} + \text{H}_2\text{CO} + (\text{NC})_2 + \text{P}_i \longrightarrow \text{Rib-1-P} + \text{NCCONH}_2 \qquad [8.11]$$

The reactions between purine bases and ribose phosphate are

$$\text{Rib-1-P} + \text{Gua} \longrightarrow \text{G} + \text{P}_i \qquad [8.12]$$

$$\text{Rib-1-P} + \text{Ade} \longrightarrow \text{A} + \text{P}_i \qquad [8.13]$$

Thus the overall equation of adenosine synthesis is obtained by the summation of Eqs. [8.10], [8.11], and [8.13]:

$$\underline{\text{P}_i} + \underline{4\text{NH}_3} + 5\text{HCN} + 2\text{GlAD} + \text{H}_2\text{CO} + (\text{NC})_2 \xrightarrow{\text{①}}$$

$$\text{A} + \underline{\text{P}_i} + \underline{4\text{NH}_3} + \text{NCCONH}_2 \qquad [8.14]$$

The overall equation of guanosine synthesis is obtained in a similar way by the summation of Eqs. [8.9], [8.11], and [8.12]:

$$\underline{3NH_3} + \underline{HCN} + \underline{P_i} + 3HCN + 2(NC)_2 + H_2O + GlAD + H_2CO$$

$$\xrightarrow{\textcircled{1}} G + \underline{3NH_3} + \underline{HCN} + \underline{P_i} + NCCONH_2 \qquad [8.15]$$

Pyrimidine bases are synthesized directly on ribose:

$$Rib + NH_2CN \longrightarrow C^1 \qquad [8.16.1]$$

$$C^1 + HC_3N \rightleftarrows C^2 \qquad [8.16.2]$$

$$C^2 \longrightarrow C^3 \qquad [8.16.3]$$

$$C^3 + H_2O \longrightarrow C \qquad [8.16.4]$$

$$Rib + NH_3CN + H_2O + HC_4N \longrightarrow C \qquad [8.16]$$

where $C^1 - C^3$ are intermediates.

The overall equation of the synthetic pathway for cytidine is obtained by the summation of Eqs. [8.16], [8.11.1], and [8.11.2]:

$$2GlAD + H_3CO + NH_2CN + HC_4N + H_2O \longrightarrow C \qquad [8.17]$$

The first two steps for the synthesis of uridine are identical with steps [8.16.1] and [8.16.2], and then proceed as follows:

$$C^2 + H_2O \longrightarrow U^3 \qquad [8.18.1]$$

$$U^3 \rightleftarrows U^4 + NH_3 \qquad [8.18.2]$$

$$U^4 + H_2O \longrightarrow U \qquad [8.18.3]$$

$$C^2 + 2H_2O \longrightarrow U + NH_3 \qquad [8.18]$$

Thus the summation of Eqs. [8.11.1], [8.11.2], [8.16.1], [8.16.2] and [8.18] leads to the overall equation of the complete synthetic pathway:

$$2GlAD + H_2CO + NH_2CN + HC_3N + 2H_2O \longrightarrow U + NH_3 \qquad [8.19]$$

The elaboration of the overall equation for the synthesis of nucleotide imidazolides is quite complicated partly because it contains a cycle in which dinucleotides may be formed in 16 different variants (i.e., each of the four different nucleotides may form four different dinucleotides) and this fact should be reflected in the overall equation, and partly because the quantitative relations (i.e., what amounts are formed from each of the nucleotide imidazolides) are dependent on the base composition of template RNAs (through consumption), and the overall equation should provide a possibility also for treating these relations quantitatively. Therefore, let us introduce the following designations:

$$AMP \equiv V_1 \qquad [8.20.1]$$

$$GMP \equiv V_2 \qquad [8.20.2]$$

$$UMP \equiv V_3 \qquad [8.20.3]$$

$$CMP \equiv V_4 \qquad [8.20.4]$$

Let us then transform reaction pathways shown in Fig.8.6 to a graph illustrated in Fig. 8.16 by applying these designations, where V_i and $V_{i'}$ are each one of the four nucleotides (i = 1,2,3, and 4).

Let us first examine the stoichiometry of the cycle:

$$V_{i'} + H_2NCN \longrightarrow V^*{}_{i'} \qquad [8.21.1]$$

$$V^*{}_{i'} + V_i \longrightarrow V_iV_{i'} + CO(NH_3)_3 \qquad [8.21.2]$$

$$V_iV_{i'} + HCONH_3 \longrightarrow V_{i'} + I_{i1} \qquad [8.21.3]$$

$$V_i + \underline{V_{i'}} + H_2NCN_2 + HCONH_2 \xrightarrow{\;①\;} \underline{V_{i'}} + I_{i1} + CO(NH_2)_2 \qquad [8.21]$$

However, the catalytic $V_{i'}$ role can be played not only by each nucleotide, but also by their arbitrary mixtures. Taking this into account, Eq. [8.21] should be modified to:

$$\sum_{i'=1}^{4} \nu_{V_{i'}} V_{i'} + V_i + H_2NCN + HCONH_2 \xrightarrow{①}$$

$$\sum_{i'=1}^{4} \nu_{V_{i'}} V_{i'} + I_{i1} + CO(NH_2)_2 \qquad [8.22]$$

where it holds that

$$\sum_{i'=1}^{4} \nu_{V_{i'}} = 1 \qquad [8.23.1]$$

but

$$0 \geq \nu_{V_{i'}} \leq I \qquad [8.23.2]$$

The steps of the linear part of the synthetic pathway are:

$$Ns_i + AcP \longrightarrow V_i + Ac \qquad [8.24.1]$$

$$I_{i1} + H_2CO \longrightarrow I_{i2} \qquad [8.24.2]$$

$$I_{i2} + NH_3 \longrightarrow I_{i3} \qquad [8.24.3]$$

$$I_{i3} + H_2CO \longrightarrow I_{i4} \qquad [8.24.4]$$

$$I_{i4} \longrightarrow V_i' \qquad [8.24.5]$$

$$Ns_i + AcP + I_{i1} + 2H_2CO + NH_3 \longrightarrow V_i + Ac + V_i' \qquad [8.24]$$

Where the I_{iz}- s are the intermediates formed in subsequent reactions steps (z= 1, 2, 3, 4) and V'_i is the nucleotide imidazolide.

By the summation of Eqs. [8.22] and [8.24] we obtain the overall equation of the synthetic pathway shown in Fig. 8.16:

$$Ns_i + \underline{\sum_{i'=1}^{4} \nu_{V_{i'}} V_{i'}} + AcP + 2H_2CO + NH_3 + H_2NCN + HCONH_2$$

$$\xrightarrow{\;①\;} V_i' + \underline{\sum_{i'=1}^{4} \nu_{V_{i'}} V_{i'}} + Ac + CO(NH_2)_2 \qquad [8.25]$$

In order to simplify the following, increasingly complex equations, let us introduce the designation

$$\sum_{i'=1}^{4} \nu_{V_{i'}} V_{i'} = W_{i'} \qquad [8.26]$$

The overall equation of the synthesis for adenine nucleotide imidazolide (i = 1) is given by the sum of Eqs. [8.14] and [8.25] considering also Eq. [8.26]:

$$\underline{W}_1 + \underline{P}_i + 4\underline{NH}_3 + 5HCN + 2GlAD + 3H_2CO + (NC)_2 + AcP + NH_3 +$$

$$+ H_2NCN + HCONH_2 \xrightarrow{\;①\;} \underline{W}_i + \underline{P}_i + 4\underline{HN}_3 + Ac + CO(NH_2)_2 +$$

$$+ NCONH_3 + V_i' \qquad [8.27]$$

that for quanine nucleotide imidazolide by the sum of Eqs. [8.15] and [8.25] by applying Eq. [8.26]:

$$\underline{W}_2 + 3\underline{NH}_3 + \underline{HCN} + \underline{P}_i + 3H_2CO + NH_3 + H_2NCN + HCONH_2 +$$

$$+ AcP + 3HCN + 2(NC)_2 + H_2O + GlAD \xrightarrow{\;①\;} V_2' + \underline{W}_2 +$$

$$+ \underline{HCN} + \underline{P}_i + 3\underline{NH}_3 + NCCONH_2 + Ac + CO(NH_2)_2 \qquad [8.28]$$

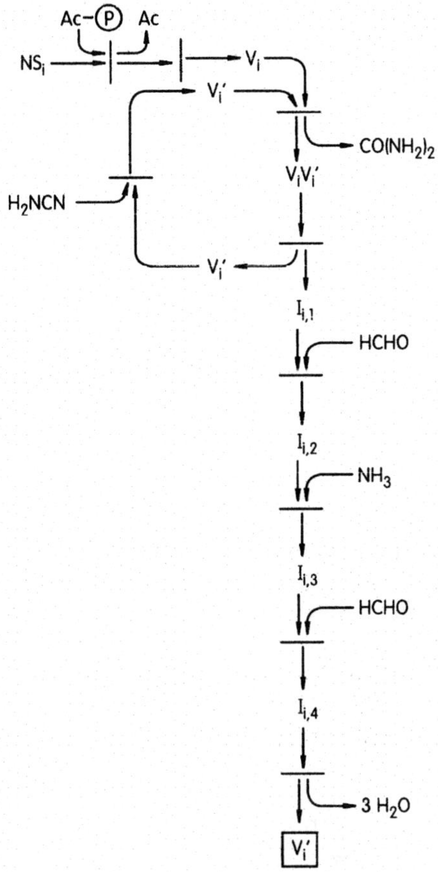

Fig. 8.16.The calculation graph of the reaction pathway shown in Fig. 8.6.

that for uracyl nucleotide imidazoide (i = 3) by summing Eqs. [8.19] and [8.25] in a similar way:

$$\underline{W}_3 + 2\text{GlAD} + 3\text{H}_2\text{CO} + 2\text{NH}_2\text{CN} + \text{HC}_3\text{N} + 2\text{H}_2\text{O} + \text{AcP} +$$

$$+ \text{HCONH}_2 \xrightarrow{\;①\;} \text{V}_3' + \underline{W}_3 + \text{Ac} + \text{CO(NH}_2)_2 \qquad [8.29]$$

and finally that for cytosine nucleotide imidazoide (i=4) by adding Eqs. [8.17] and [8.25] together and considering also Eq. [8.26]:

$$\underline{W}_4 + 2GlAD + 3H_2CO + 2NH_2CN + HC_3N + 2H_2O + AcP +$$

$$+ HCONH_2 \xrightarrow{①} V_4' + \underline{W}_4 + Ac + CO(NH_2)_2 \qquad [8.30]$$

It is obvious that in these equations

$$V_1' \equiv AMP - imidazolide \qquad [8.31.1]$$

$$V_2' \equiv GMP - imidazolide \qquad [8.31.2]$$

$$V_3' \equiv UMP - imidazolide \qquad [8.31.3]$$

$$V_4' \equiv CMP - imidazolide \qquad [8.31.4]$$

In the possession of the overall equations describing the formation of the building units of the template, our next task is to learn what amounts of each are necessary for the replication of the template RNA.

The general equation for the need for building units in template polymerization is given in Eq. [2.147] (Gánti, 1978a, 1979a). The application of this equation for the RNA template of prebiotic chemotons yields:

$$\sum_{i=1}^{4} \kappa_i V_i' + \sum_{i=1}^{4} \underline{p\kappa_i V_i} \xrightarrow{①} 2\sum_{i=1}^{4} \underline{p\kappa_i V_i} + \left(\sum_{i=1}^{4} \kappa_i - 2 \right) R \qquad [8.32]$$

with the number of building units in RNA being

$$k = 4 \qquad [8.33]$$

As in Eq. [8.32], κ_i refers to the relative amounts of individual nucleotides, and as according to Chargaff's rule, in the double helix of RNA the molar amounts of A and U, as well as those of G and C are identical (as a consequence of pair formation), it holds that

$$\kappa_1 = \kappa_3 \qquad [8.34.1]$$

$$\kappa_2 = \kappa_4 \qquad [8.34.2]$$

At the right of Eq. [8.32], -2 in the multiplication factor preceding R shows that in the RNA double helix R (imidazole) remains fixed on the last nucleotides of both strands.

Thus the complete overall equation for the polycondensation of the RNA template is obtained if Eqs. [8.27] and [8.29] describing the synthesis of V_1' and V_2' arc multiplicd by κ_1, Eqs. [8.28] and [8.30] charactcrizing thc formation of V_2 and V_4 are multiplied by κ_2, and these and Eq. [8.32] are added together:

$$\underline{\sum_{i=1}^{4} p\kappa_i V_i} + \underline{\sum_{i=1}^{4} \kappa_i W_i} + \left(\sum_{i=1}^{4} \kappa_i + 2\kappa_1 + \kappa_2\right)\underline{NH_3} + \sum_{i=1}^{2} \kappa_i P +$$

$$+ \kappa_2 \underline{HCN} + 2\sum_{i=1}^{4} \kappa_i GlAD + \left(\kappa_2 + \sum_{i=1}^{4} \kappa_i\right)(NC)_2 + \sum_{i=1}^{4} \kappa_i AcP +$$

$$+ 1{,}5\sum_{i=1}^{4} \kappa_i NH_2CN + \sum_{i=1}^{2} \kappa_i HC_3N + \sum_{i=1}^{4} \kappa_i H_2O + \left(\kappa_2 + \sum_{i=1}^{2} \kappa_i\right) NH_3 +$$

$$+ \left(1{,}5\sum_{i=1}^{4} \kappa_i + 2\kappa_1\right) HCN + 3\sum_{i=1}^{4} \kappa_i H_2CO + \sum_{i=1}^{4} \kappa_i HCONH_2 \xrightarrow{\text{①}}$$

$$2\underline{\sum_{i-1}^{4} p\kappa_i V_i} + \sum_{i-1}^{4} \kappa_i \underline{W}_i + (\kappa_1 + 1{,}5)\sum_{i-1}^{4} \kappa_i \underline{NH_3} + \sum_{i-1}^{2} \kappa_i \underline{P} + \kappa_2 \underline{HCN} +$$

$$+ \sum_{i=1}^{4} \kappa_i CO(NH_2)_2 + \sum_{i=1}^{2} \kappa_i NCCONH_2 + \left(\sum_{i=2}^{4} \kappa_i - 2\right) R \qquad [8.35]$$

However, nucleotides are necessary not only for the replication of the template in the metabolism of prebiotic chemotons. The amounts of nucleotides playing a catalytic role as cycle intermediates in Figs. 8.6 and 8.16 should also be duplicated in the period between two divisions of the chemoton. The equations of synthetic pathways describing this process are the following:

In the case of $i' = 1$, Eqs. [8.14] and [8.24.1] should be added together:

$$\underline{P} + 4\underline{NH_3} + 5HCN + 2GlAD + (NC)_2 + H_2CO + AcP \xrightarrow{①}$$

$$V_1 + Ac + 4\underline{NH_3} + \underline{P} + NCONH_2 \qquad [8.36]$$

If $i' = 2$, the synthetic pathway is obtained by the addition of Eqs. [8.15] and [8.24.1]:

$$3\underline{NH_3} + \underline{P} + \underline{HCN} + 3HCN + 2(NC)_2 + H_2O + 2GlAD + H_2CO +$$

$$+ AcP \xrightarrow{①} V_2 + 3\underline{NH_3} + \underline{P} + \underline{HCN} + NCCONH_2 + Ac \qquad [8.37]$$

If i'= 3, the equation is derived by the summation of Eqs. [8.19] and [8.24.1]:

$$2GlAD + H_2CO + NH_2CN + HC_3N + AcP \longrightarrow V_3 + NH_3 + Ac \qquad [8.38]$$

Finally, in the case of i'=4, Eqs. [8.17] and [8.24.1] should be added together:

$$2GlAD + H_2CO + NH_2CN + HC_3N + H_2O + AcP \longrightarrow V_4 + Ac \qquad [8.39]$$

It should be remembered that V_i is an abbreviation (see Eq. [8.26]), thus the expression in Eq. [8.35] is in its full form

$$\sum_{i=1}^{4} \kappa_i W_i \equiv \sum_{i=1}^{4} \kappa_i \sum_{i'=1}^{4} \nu_{V_{i'}} V_{i'} \qquad [8.40]$$

Let us introduce the following designation:

$$\sum_{i=1}^{4} \kappa_i \nu_{V_{i'}} = \nu_j \quad (j = 1,2,3,4) \qquad [8.41]$$

The overall equation describing the duplication of the amounts of nucleotides playing the catalytic role can be given by multiplying Eqs. [8.36] – [8.39] by the corresponding v_j -s and then summing them up:

$$\sum_{j=1}^{2} v_j P + (4v_1 + 3v_2)\underline{NH_3} + v_2\underline{HCN} + 2\sum_{j=1}^{4} v_j GlAD + \sum_{j=1}^{4} v_{j_i} AcP +$$

$$+ \sum_{j=1}^{4} v_j H_2CO + (v_1 + 2v_2)(NC)_2 + (5v_1 + 3v_2)HCN + (v_2 + 2v_3 + v_4)H_2O +$$

$$+ \sum_{j=3}^{4} v_j NH_2CN + \sum_{j=3}^{4} v_j HC_3N \xrightarrow{\text{①}}$$

$$\sum_{i=1}^{4} \kappa_i W_i + \sum_{j=1}^{2} v_j P + (4v_1 + 3v_2)\underline{NH_3} +$$

$$+ v_2\underline{HCN} + \sum_{j=1}^{2} v_j NCCONH_2 + \sum_{j=1}^{4} v_j Ac + v_3 NH_3 \qquad [8.42]$$

Combination of the two subsystems

The overall equation for the coupled operation of the membrane and template subsystems is provided by the summation of overall equations [8.8], [8.35], and [8.42]:

$$
\begin{aligned}
&\sum_{i=1}^{4} \underline{p\kappa_i V_i} + \sum_{i=1}^{4} \underline{\kappa_i W_i +}\ \boxed{T_m} + [4(\kappa_1+\nu_1)+3(\kappa_2+\nu_2)]\underline{NH_3} + \\
&+(\kappa_2+\nu_2)\underline{HCN} + \left(\sum_{i=1}^{2}\kappa_i+\sum_{j=1}^{2}\nu_j\right)P_i + 2\left(\sum_{i=1}^{4}\kappa_i+\sum_{j=1}^{4}\nu_j\right)GlAD + \\
&+[\kappa_1+\nu_1+2(\kappa_2+\nu_2)](NC)_2 + \left(\sum_{i=1}^{4}\kappa_i+\sum_{j=1}^{4}\nu_j+48m\right)AcP + \\
&+\left(3\sum_{i=1}^{2}\kappa_i+\sum_{j=3}^{4}\nu_j\right)NH_2CN + \left(\sum_{i=1}^{2}\kappa_i+\sum_{j=3}^{4}\nu_j\right)HC_3N + \\
&+\left(\sum_{i=1}^{4}\kappa_i+\sum_{j=2}^{4}\nu_j+\nu_3\right)H_2O + \left(\sum_{i=1}^{2}\kappa_i+\kappa_2-\nu_3\right)NH_3 + \\
&+[5(\kappa_1+\nu_1)+3(\kappa_2+\nu_2)]HCN + \left(3\sum_{i=1}^{4}\kappa_i+\sum_{j=1}^{4}\nu_j\right)H_2CO + \\
&+\sum_{i=1}^{4}\kappa_i HCONH_2 + mGADP + m48H* \xrightarrow{\textcircled{1}} \\
&2\left(\sum_{i=1}^{4}\underline{p\kappa_i V_i} + \sum_{i=1}^{4}\underline{\kappa_i W}\right) + \boxed{T_{2m}} + 4[(\kappa_1+\nu_1)+3(\kappa_2+\nu_2)]NH_3 + \\
&+(\kappa_2+\nu_2)HCN + \left(\sum_{i=1}^{2}\kappa_i+\sum_{j=1}^{2}\nu_j\right)P_i + \left(\sum_{i=1}^{4}\kappa_i+\sum_{j=1}^{4}\nu_j+m24\right)Ac + \\
&+\left(\sum_{i=1}^{2}\kappa_i+\sum_{j=1}^{2}\nu_j\right)HCCONH_2 + \sum_{i=1}^{4}\kappa_i CO(NH_2)_2 + m49P_i + m8CO_2 \qquad [8.43]
\end{aligned}
$$

where

$$\sum_{i=1}^{1} \kappa_i - 2 = m \qquad [8.44]$$

which is the condition for a stoichiometric coupling of the membrane and template systems through imidazole.

Synthesis of precursors

Among the components on the left side of Eq. [8.43], neither glyceraldehyde phosphate nor acetylphosphate may be considered nutrients from outside, because they cannot be taken up, thus they should also be prepared inside the chemoton.

The formation of GADP is a simple process:

$$GlAD + H_2CO \rightleftharpoons GAD \qquad [8.45.1]$$

$$GAD + AcP \longrightarrow GADP + Ac \qquad [8.45.2]$$

$$GlAD + H_2CO + AcP \longrightarrow GADP + Ac \qquad [8.45]$$

The necessary amount of GADP according to Eq. [8,43] is obtained upon multiplication by m:

$$mGlAD + mH_2CO + mAcP \longrightarrow mGADP + mAc \qquad [8.46]$$

As mentioned earlier, acetylphosphate acts as both phosphorylating and acetylating agent. In phosphorylation, it has a coenzyme function, AcP consumed is regenerated continuously, i.e., only part of the phosphorylating AcP used up for growth and proliferation should be synthesized *de novo*. However, the part of AcP utilized for acetylation has to be completely synthesized.

The *de novo* synthesis of AcP occurs in a cycle, as is shown in Fig. 8.13:

$$\text{GADP} \rightleftarrows (\text{OH})_2-\text{PON}-\text{P} \qquad [8.47.1]$$

$$(\text{OH})_2-\text{PON}-\text{P}+\text{H}_2\text{CO} \rightleftarrows \text{2-TuP} \qquad [8.47.2]$$

$$\text{2-TuP}+\text{H}_2\text{CO} \rightleftarrows \text{3-PE}-\text{P} \qquad [8.47.3]$$

$$\text{3-PE}-\text{P} \rightleftarrows \text{XuP} \qquad [8.47.4]$$

$$\text{XuP}+\text{P}_\text{i} \rightleftarrows \text{AcP}+\text{GADP} \qquad [8.47.5]$$

$$\underline{\text{GADP}}+2\text{H}_2\text{CO}+\text{P}_\text{i} \xrightarrow{①} \underline{\text{GADP}}+\text{AcP} \qquad [8.47]$$

Numerous cycles are involved already in the network discussed so far. As was shown earlier, the coupling of cycles can only be treated in terms of stoichiokinetics in a fully general form. In the equations derived so far, the use of stoichiokinetics has been avoided, in order not to make the complex stoichiometric description of the system even more complicated by providing the completely general equations. In other words, in all processes where the amount of intermediates can be arbitrarily varied by changing the turning numbers of cycles, we have taken into consideration the stoichiometric amounts that ensure the necessary amount of products in one turn of the cycle. This is a chemically faultless solution, but not the only one; numerous other stoichiometric variants are also possible.

Nevertheless, in describing the regeneration process of acetylphosphate, the application of stoichiokinetics cannot be avoided any more. In the process characterized by Eq. [8.47], namely, an amount of AcP equivalent to that of GADP is formed. However, until the division of the chemoton, the amount of internal substances, among them also GADP should be doubled. The newly synthesized GAD in our network is phosphorylated by AcP, thus in this process the *de novo* synthesized AcP is quantitatively consumed. On the other hand, the AcP formed (regenerated) in one regeneration cycle cannot satisfy the demands for phosphorylation and acetylation processes in the other parts of the network, therefore, the available AcP should be regenerated in more cycles, and this is why the quantitative treatment is only possible by means of stoichiokinetics.

The AcP cycle (regeneration and consumption) may be written in its general form as follows:

$$AcP + X \longrightarrow Ac + XP \qquad [8.48.1]$$

$$Ac + P_i \longrightarrow AcP \qquad [8.48.2]$$

$$\underline{AcP} + X + P_i \xrightarrow{\quad ① \quad} \underline{AcP} + XP \qquad [8.48]$$

where X refers to the actual phosphate acceptor in the individual processes.

Equation [8.48.2] is a generalized scheme; it may be based, e.g., on the hypothesis that Halmann's reaction can be applied also for the formation of acetylphosphate, i.e. in the presence of cyanoderivatives (perhaps via cyanovinylphosphate) Ac is phosphorylated to AcP.

Equation [8.48] can be generalized by introducing the arbitrary turning number u and the stoichiometric factor ν_{AcP}

$$\nu_{\underline{AcP}}\underline{AcP} + u\nu_{\underline{AcP}}X + \nu u_{\underline{AcP}}P_i \xrightarrow{\quad ① \quad} \nu_{\underline{AcP}}\underline{AcP} + \nu u_{\underline{AcP}}XP \qquad [8.49]$$

which may also be written in the following form

$$\nu_{\underline{AcP}}\underline{AcP} + \mu_{\underline{AcP}}X + \mu_{\underline{AcP}}P_i \xrightarrow{\quad ① \quad} \nu_{\underline{AcP}}\underline{AcP} + \mu_{\underline{AcP}}XP \qquad [8.50]$$

where it holds that

$$\mu_{\underline{AcP}} = u\nu_{\underline{AcP}} \qquad [8.51]$$

μ_{AcP} is the stoichiokinetic coefficient related to the AcP cycle.

In order to obtain the part of Eq. [8.47] providing the amount of phosphorylating AcP, Eq. [8.47] should be multiplied by the stoichiometric factor ν_{AcP}:

$$\nu_{\underline{AcP}}\underline{GADP} + \nu_{\underline{AcP}}2H_2O + \nu_{\underline{AcP}}P_i \xrightarrow{\quad ① \quad}$$

$$\nu_{\underline{AcP}}\underline{GADP} + \nu_{\underline{AcP}}\underline{AcP} \qquad [8.52]$$

On the other hand, a significant amount of AcP is used for acetylation; this part is obtained by the subtraction of Ac formed from the total amount of AcP consumed in the overall equation [8.43]:

$$\sum_{i=1}^{4}\kappa_i + \sum_{j=1}^{4}\nu_j + 48m - \sum_{i=1}^{4}\kappa_i - \sum_{J=1}^{4}\nu_j - 24m = 24m \qquad [8.53]$$

The total demand of *de novo* synthesized AcP is then obtained by the summation of Eqs. [8.53] and [8.53]:

$$(24m + \nu_{\underline{AcP}})\underline{GADP} + 2(24m + \nu_{\underline{AcP}})H_2CO + (24m + \nu_{\underline{AcP}}P_i)$$

$$\xrightarrow{\;①\;} (24m + \nu_{\underline{AcP}})\underline{GADP} + (24m + \nu_{\underline{AcP}})AcP \qquad [8.54]$$

The duplication of the amount of GADP also requires AcP:

$$(25m + \nu_{\underline{AcP}})GlAD + (25m + \nu_{\underline{AcP}})H_2CO + (25m + \nu_{\underline{AcP}}) \longrightarrow$$

$$(25m + \nu_{\underline{AcP}})GADP + (25m + \nu_{\underline{AcP}})Ac \qquad [8.55]$$

For the calculation of Eq. [8.50], the amount of AcP used for phosphorylation and given by Eq. [8.43] should be considered together with the amount in Eq. [8.55]:

$$\mu_{\underline{AcP}} = 49 + \sum_{i=1}^{4}\kappa_i + \sum_{i=1}^{4}\nu_i + \nu_{\underline{AcP}} \qquad [8.56]$$

Self-reproducing metabolic network

In the prebiotic chemoton described here, the metabolic network capable of self-production is the autocatalytic formose network. Its reaction steps are the following:

$$\mathrm{GlAD} + \mathrm{H_2CO} \rightleftarrows \mathrm{GAD} \qquad [8.57.1]$$

$$\mathrm{GAD} \rightleftarrows \mathrm{(OH)_2\text{-}PON} \qquad [8.57.2]$$

$$\mathrm{(OH)_2\text{-}PON} + \mathrm{H_2CO} \rightleftarrows \mathrm{Tu} \qquad [8.57.3]$$

$$\mathrm{Tu} + \mathrm{H_2CO} \rightleftarrows \mathrm{PE} \qquad [8.57.4]$$

$$\mathrm{PE} \rightleftarrows \mathrm{Xu} \qquad [8.57.5]$$

$$\mathrm{Xu} + \mathrm{H_2CO} \rightleftarrows \mathrm{3\text{-}Hu} \qquad [8.57.6]$$

$$\mathrm{3\text{-}Hu} \rightleftarrows \mathrm{2\text{-}Hu} \qquad [8.57.7]$$

$$\mathrm{2\text{-}Hu} \rightleftarrows \mathrm{AHex} \qquad [8.57.8]$$

$$\mathrm{AHex} \rightleftarrows \mathrm{ATet} + \mathrm{GlAD} \qquad [8.57.9]$$

$$\mathrm{ATet} \rightleftarrows \mathrm{2GlAD} \qquad [8.57.10]$$

$$\underline{\mathrm{GlAD}} + \mathrm{4H_2CO} \xrightarrow{(1)} 3\underline{\mathrm{GlAD}} \qquad [8.57]$$

The advantage of the description of the formose network in this form is that the amount of the starting material (GlAD) is not only doubled but tripled in one cycle, which makes it possible to let the cycle operate without tapping, i.e., upon using half of the products for other synthesis routes, the self-reproduction of the cycle is still ensured by the other half.

On arranging Eq. [8.57] so that it satisfies the quantitative demands of other reaction pathways of the prebiotic chemoton we obtain:

$$\nu_{\underline{GlAD}}\underline{GlAD} + 4\nu_{\underline{GlAD}}H_2CO \xrightarrow{①}$$

$$\nu_{\underline{GlAD}}\underline{GlAD} + 2\nu_{\underline{GlAD}}\underline{GlAD} \qquad [8.58]$$

where stoichiometric coefficient ν_{GlAD} determines the requirements of the prebiotic chemoton network for GlAD, it can be calculated from Eqs. [8.43] and [8.55]:

$$\nu_{\underline{GlAD}} = 2\left(\sum_{i=1}^{4}\kappa_i + \sum_{j=1}^{4}\nu_j\right) + 25m + \nu_{\underline{AcP}} \qquad [8.59]$$

Overall equation of the prebiotic chemoton

Overall equations [8.43], [8.50], [8.54], [8.55], and [8.58] provide the quantitative relations of all sequences of chemical events taking place in prebiotic chemotons. The only exceptions are the parts of NH_3, HCN, and P_i that are not only consumed as initial materials, but are also formed in the reactions, i.e., that have been regarded as NH_3, *HCN,* and P_i in the stoichiometric discussions – components regenerating cyclically. These amounts may be considered simply by writing them at both sides of the overall equation as cyclic components multiplied by the appropriate stoichiometric factors.

Hence the overall equation of prebiotic chemotons can be derived by the summation of cyclic overall equations [8.43], [8.50], [8.54], [8.55], and [8.58] including the cyclically regenerating parts of and as cyclic components:

$$\left[\nu_{\underline{GlAD}}\underline{GlAD} + \nu_{\underline{GADP}}\underline{GADP} + \mu_{\underline{AcP}}\underline{AcP} + \sum_{i=1}^{4}\underline{p\kappa_i V_i} + \sum_{i=1}^{4}\underline{\kappa_i W_i} + \boxed{T_m}\right.$$

$$\left. + \nu_{\underline{NH}}\underline{NH_3} + \nu_{\underline{HCN}}\underline{HCN} + \nu_{P_i}\underline{P_i}\right] + \nu_{(NC)_2}(NC)_2 + \nu_{NH_2CN}NH_2CN +$$

$$+ \nu_{HC_4N}HC_4N + \nu_{H_2O}H_2O + \nu_{H_2CO}H_2CO + \nu_{HCN}HCN + \nu_{NH_4}NH_4 + \nu_{P_i}P_i +$$

$$+ \nu_{HCONH_2}HCONH_2 + \nu_{H^*}H^* \xrightarrow[\text{①}]{\text{Prebiotic chemoton}}$$

$$2\left[\nu_{\underline{GlAD}}\text{GlAD} + \nu_{\underline{GADP}}\text{GADP} + \mu_{\underline{AcP}}\text{AcP} + \sum_{i=1}^{4} \underline{p\kappa_i V_i} + \sum_{i=1}^{4} \underline{\kappa_i W_i} + \right.$$

$$\left. + \boxed{T_m} + \nu_{\underline{NH_3}}\underline{NH_3} + \nu_{\underline{HCN}}\underline{HCN} + \nu_{\underline{P_i}}\underline{P_i}\right] + \nu_{NCCONH_2}NCCONH_2 +$$

$$+ \nu_{CO(NH_2)_2}CO(NH_2)_2 + \nu_{CO_2}CO_2 \qquad [8.60]$$

where the members in brackets on the left side of the equation represent the composition of the initial chemoton, the subsequent ones are the nutrients consumed for the first division, whereas on the right side of the equation the composition of the two chemotons after division is shown in brackets, followed by the members representing the waste materials. The values of multiplication factors in the equation are:

from Eqs. [8.44] and [8.59]:

$$\nu_{\underline{GlAD}} = 27\sum_{i=1}^{4}\kappa_i - 50 + 2\sum_{i=1}^{4}\nu_i + \nu_{\underline{AcP}} \qquad [8.61.1]$$

from Eqs. [8.44] and [8.54]:

$$\nu_{\underline{GADP}} = 24\sum_{i=1}^{4}\kappa_i - 48 + \nu_{\underline{AcP}} \qquad [8.61.2]$$

from Eqs. [8.44] and [8.56]:

$$\mu_{\underline{AcP}} = 50\sum_{i=1}^{4}\kappa_i - 98 + \sum_{i=1}^{4}\nu_i + \nu_{\underline{AcP}} \qquad [8.61.3]$$

from Eq. [8.43]:

$$\nu_{\underline{NH}} = 4(\kappa_1 + \nu_1) + 3(\kappa_2 + \nu_2) \qquad [8.61.4]$$

from Eq. [8.43]:

$$\nu_{\underline{HCN}} = \nu_2 + \kappa_2 \qquad [8.61.5]$$

from Eqs. [8.43],[8.44], and [8.50]:

$$\nu_{\underline{P}_i} = 49{,}5\sum_{i=1}^{4}\kappa_i - 98 + \sum_{j=1}^{2}\nu_j \qquad [8.61.6]$$

from Eq. [8.43]:

$$\nu_{(NC)_2} = \kappa_1 + \nu_1 + 2(\kappa_2 + \nu_2) \qquad [8.61.7]$$

from Eq. [8.43]:

$$\nu_{NH_2CN} = 1{,}5\sum_{i=1}^{4}\kappa_i + \sum_{j=3}^{4}\nu_j \qquad [8.61.8]$$

from Eq. [8.43]:

$$\nu_{HC_3N} = 0{,}5\sum_{i=1}^{4}\kappa_i + \sum_{j=3}^{4}\nu_j \qquad [8.61.9]$$

from Eqs.[8.43], [8.44], [8.54], [8.55], [8.58], and [8.61]:

$$\nu_{H_2O} = \sum_{i=1}^{4}\kappa_i + \sum_{j=2}^{4}\nu_j + \nu_3 \qquad [8.61.10]$$

from Eq. [8.43]:

$$\nu_{H_2CO} = 184\sum_{i=1}^{4}\kappa_i + 9\sum_{j=1}^{4}\nu_j - 346 + 7\nu_{\underline{AcP}} \qquad [8.61.11]$$

from Eq. [8.43]:

$$\nu_{HCN} = 5(\kappa_1 + \nu_1) + 3(\kappa_2 + \nu_2) \qquad [8.61.12]$$

from Eq. [8.43]:

$$\nu_{NH_3} = 2{,}5\sum_{i=1}^{4}\kappa_i + 4\nu_1 + 3\nu_2 - \nu_3 \qquad [8.61.13]$$

from Eqs. [8.43], [8.44], [8.50], [8.54], [8.61.3], and [8.61.6]:

$$\nu_P = 74{,}5\sum_{i=1}^{4}\kappa_i + \sum_{j=1}^{4}\nu_j + 2\nu_{\underline{AcP}} - 146 + \sum_{j=1}^{3}\nu_j \qquad [8.61.14]$$

$$\nu_{HCONH_2} = \sum_{i=1}^{4}\kappa_i \qquad [8.61.15]$$

from Eqs. [8.43] and [8.44]:

$$\nu_{H^{\bullet}} = 48\sum_{i=1}^{4}\kappa_i - 96 \qquad [8.61.16]$$

from Eq. [8.43]:

$$\nu_{NCCONH_2} = 0{,}5\sum_{i=1}^{4}\kappa_i + \sum_{j=1}^{2}\nu_j \qquad [8.61.17]$$

from Eq. [8.43]:

$$\nu_{CO(NH_2)_2} = \sum_{i=1}^{4}\kappa_i \qquad [8.61.18]$$

and finally from Eqs. [8.43] and [8.44]:

$$\nu_{CO_2} = 8\sum_{i=1}^{4}\kappa_i - 16 \qquad [8.61.19]$$

IX. Prebiotic Evolution

Introduction

In the introductory part of Chapter VI an objective was set: to produce basic principles for an exact, quantitative theoretical biology that would enable us to derive both qualitatively and quantitatively the many-sided, complex phenomena of the living world from the basic principles of biology. This is the reason why in Chapter VI the most general common properties of living systems were summarized and a system of criteria designed, according to which the system in question may be qualified as a living one. Further it has been established that if this system of criteria is accepted for telling living and nonliving systems apart, then the proliferating program-controlled fluid automata (i.e., chemotons as described in Chapter IV) should be regarded as living. More precisely: all systems, the organization of which satisfies the criteria of chemotonial organization as provided by Eq. [4.67], should be considered living. It has also been proven that chemotons are biological minimum systems, i.e., no system with a simpler stoichiometry can be constructed that would satisfy the life criteria. However, it has not been stated that this is the only organizational mode satisfying life criteria.

Chemotons – as is shown in Chapter IV – can be treated in a quantitatively exact manner by means of cycle stoichiometry. Thus they seem to be suitable as a starting basis for a quantitative theoretical biology. However, in the form shown in Chapter IV, they are only abstract constructions. Therefore, in Chapter VIII, it was proved that such systems could have formed and proliferated under primitive conditions during chemical evolution, that the total metabolic network of the system can also be given in detail, and that the operation of these systems can be described quantitatively by the cycle stoichiometry (and stoichiokinetics) elaborated for the characterization of chemotons.

The concrete metabolic network (shown as an example) illustrates the probability that, if such systems of chemotonial organization formed spontaneously under primordial conditions, then from those systems our present living world could have developed. This is so because, first, they are based on a carbohydrate metabolism similar to present living beings; second, their program control is based on nucleic acid template polymerization; and finally, the construction of their membrane is analogous to that of archaebacteria. However, there are also basic differences between prebiotic chemotons and the simplest prokaryotic systems. In the living world – and thus also in the simplest prokaryotes – the hereditary information is carried by the nucleotide sequence of nucleic acids, whereas in the simple prebiotic chemotons, only by the composition of nucleic acids. Also in the living world, the controlling program is carried by DNA and in prebiotic chemotons by RNA. Finally, the systems of the living world, in addition to their being controlled stoichiometrically and kinetically, have catalytic regulation through their enzyme proteins, whereas prebiotic chemotons do not have such a catalytic regulation.

Regardless of how much the mechanisms of biological evolution are currently debated, the process reaching from prokaryotes to humans can itself be interpreted quite satisfactorily. In the molecular background of evolution the sequences of nucleic acids and proteins and/or the variations or rearrangements within these sequences have to be looked for. Obviously, these mechanisms cannot be applied for the evolution of prebiotic chemotons. If we attempt to derive the evolution of the living world from the appearance of prebiotic chemotons under primordial conditions, we have to reveal – at least in their main aspects – the possibilities that would have to be passed through by millions of generations of these systems until today's prokaryotes could develop. In other words, we have to show that hereditary changes in prebiotic chemotons could in fact lead to information storage by sequences, to enzyme activities, and to the appearance of genes carried by DNA. The objective of this chapter is to outline this possibility.

Beginnings of Prebiotic Evolution

Conditions of biogenesis on Earth

Living systems could develop on Earth only after the appearance of a liquid aqueous phase or (on the geological time scale) simultaneously to this phase.

Living beings on Earth are in fact fluid automata in which the fluid phase is water; on the other hand, they are built of organic compounds, which, under prebiotic conditions, can only be formed in an aqueous phase.

If Earth, during its formation, passed through a glowing state, liquid water could appear first in the atmosphere. Woese (1979) was the first to suppose that the genesis of life had already begun before the appearance of Primordial Ocean and surface waters in the atmosphere, in droplets constituting clouds. He argued that the hot surface of Earth generated strong air currents in the primordial atmosphere (eventually also taking along fine dust particles that contained mineralic components) and the upper regions of these currents when approaching cooler and cooler regions, allowed the formation of water drops, similarly to the formation of clouds in ascending currents that we can observe at present. However, due to the much higher temperature differences at that time, the currents were much more intense, and thus could keep larger drops in equilibrium for longer times than today's upward currents. According to Woese, these droplets – being in direct contact with gases that could provide the raw materials for metabolic processes – may be regarded as individual living beings. By this hypothesis, the compartmentalization problems of (the first) primordial living beings are circumvented.

This hypothesis of Woese seems to be acceptable, particularly if we consider the following:

1. If convections in the atmosphere originate not from solar energy (which changes at the surface of the Earth with time), but from the internal heat of the Earth (which is constant), a convection structure of constant nature is developed in the atmosphere, similar to the Benard phenomenon. Such an essentially unchanged convection structure is found in the atmosphere of Jupiter. In the compartments of the ascending air current, water droplets may be stabilized for years; thus the floating water droplets may be regarded as a constant environment with regard to the formation of life.

2. Szádeczky-Kardoss (1974) supposed that the primordial atmosphere was denser than today's atmosphere, and barometric pressure could have been as high as 160 atm. If we assume that the carbon presently found in biogenically formed carbonate rocks was present on primordial Earth in the form of volatile carbon compounds, then the atmosphere of primordial Earth must have been even denser, and the surface pressure could have been as high as 200–300 atm (Gánti, 1980). This assumption seems to be supported by the fact that the surface pressure on Venus is about 100 atm

(though Venus is smaller than the Earth) but – as is now known – there are no sedimentary carbonate rocks on its surface.

If the atmosphere of primordial Earth was 200–300 times denser than it is today, then the ascending air currents, which also had convection flows of a much higher speed, could keep much larger droplets floating than those in our present atmosphere. This could ensure extraordinarily favorable conditions for the formation of prebiotic chemotons, because the reactive compounds in the atmosphere (HCN, HCHO, NH_3, etc.) serving as nutrients for prebiotic chemotons could reach the chemotons along a very short diffusion path, and lacking sufficient time, they could not react further in the aqueous phase to form compounds for which the membranes of chemotons would be impermeable.

If we also consider that the dense masses of air in strong convection could have also transported great amounts of dust, thereby providing inorganic ions, we may conjecture that the droplets floating in the primordial atmosphere could actually be suitable places for the spontaneous genesis of living systems. However, not quite as Woese has suspected, that every droplet is a separate individuum, but rather that these droplets provided a medium in which chemotons could appear and proliferate. If the dimension of prebiotic chemotons is taken to be 1 μm, based on the experiments with microspheres, and the dimension of floating droplets in the dense atmosphere is assumed to be 1 mm, the volume of the droplets is 10^7–10^9 times that of chemotons. This volume is also more than adequate as a nutrient volume ensuring proliferation, while still being small enough for the supply of nutrients from the gaseous phase by mere diffusion.

The first hypotheses concerning biogenesis assumed that life originates from the primordial ocean (Oparin, 1960). However, this assumption seems to become more and more improbable. The water-soluble components of gases in the atmosphere, being simultaneously also the raw materials of chemical evolution, got into the sea by sediment precipitation rather than by direct dissolution. Chemical evolution experiments show that these gases react with each other very quickly in the aqueous phase, and by further combination of primary products, large molecules in a very short time form that are no longer capable of playing the role of primary nutrients in the formation of living systems. The first and also part of the second phase of reactions had probably occurred already in the water drops floating in the atmosphere, as gases could be dissolved orders of magnitude faster in the aerosol of the atmosphere than in the ocean, as a result of their very high specific surface.

Sidney Fox assumed that the cradles of life were the pools developed from precipitation falling onto the hot volcanic rocks, evaporating and consisting of caramelized organic substances. During the periods in which the Earth was cooling down, the formation of such pools was probably necessary and quite usual. Experiments carried out under similar conditions showed that from amino acids thermal proteinoids must have been formed. However, these pools could only be suitable places for the genesis of life if, in this process, thermal proteinoids played the decisive role. But even in that case, the question arises as to how primary living systems could proliferate in continuously evaporating pools.

Based on the discovery of bacteria capable of living and proliferating at 250 °C in "black smoker" types of deep sea hot springs by Baross (Baross, Lilley, and Gordon, 1982; Baross and Deming, 1983), Corliss (1984) assumed that life originated from deep-sea hot springs. The discovery by Baross has been doubted by numerous authors. For example, White (1984) showed that at 250 °C, under anaerobic conditions, the half lifetime of the various compounds with biological importance varied between minutes and hours, thus under these conditions a metabolic system consisting of usual biomolecules could not exist. On the other hand, Trent, Chastain, and Yayanos (1984) repeated Baross's experiments, and obtained similarly positive results without, however, inoculating the sterile medium with bacteria that proliferate at 250 °C! From this result they concluded that the thermally extrastable bacteria of Baross and Deming were, in fact, artifacts.

The existence of bacteria resistant to 250 °C seems to be improbable. In spite of this, however, the assumption by Corliss that life originated from deep-sea hot springs does not have to be totally rejected, at least not in the sense that life could have originated there (or could even now), if not in the regions of several hundred degrees centigrade, then perhaps in somewhat cooler regions. From this point of view, the control experiment of Trent et al. is of special importance. They observed the appearance and proliferation of compounds showing protein and nucleic acid reactions even in the non-inoculated medium under 250 °C "culture" conditions, together with the appearance of bacteria-like structures. From these data, the conclusion can be drawn that the major part of prebiotic evolutionary processes take place in hours under the given experimental conditions, though the authors themselves did not draw this conclusion. Unfortunately, the chemistry of organic compounds in the "black smokers" has not as yet been clarified enough to provide the possibility of discussing the genesis of chemotonial systems in deep-sea hot springs.

Spontaneous genesis of proliferating systems

According to the above considerations, water droplets in the primordial atmosphere seem to be the most probable site for biogenesis. Events taking place in aqueous aerosols may best be reconstructed on the basis of the experiments of Folsome and Brittain (1981). The effect of temperature and pressure may be considered negligible because under primordial atmospheric conditions, every range of temperature and pressure exists under which water is in its liquid state. From this aspect, it is of no importance if life began under warm or cold conditions.

As was mentioned, Folsome and Brittain irradiated a gaseous mixture containing N_2, CH_4, and CO_2 in the presence of an aqueous phase with 254-nm UV radiation. A great amount of microspheres was formed in the aqueous phase, having diameters from 1 μm to 20 μm. The authors found that the formation rate of microspheres increased exponentially, which means that the already-formed spheres promoted the formation of new, similar structures. However, they did not report in their paper whether this process occurred by growing and division of existing microspheres or by budding, or in any other way. Nevertheless, they established that the incorporation of carbon into the structures was the result of photoreduction and that the first stable intermediate produced was formaldehyde.

On the other hand (as discussed in Chapter VII), we know from other experiments about chemical evolution that in the presence of water, formaldehyde is formed from CH_4, CO, or CO_2 by UV irradiation, which means that if the primitive atmosphere contained carbon in the form of either CH_4, CO, or CO_2, formaldehyde could have been produced continuously by solar radiation. This formaldehyde would obviously dissolve in the water drops, thereby starting formose reactions in the drops. Thus in water drops both formaldehyde and the intermediates of the autocatalytic formose cycle were present!

As is proven by the Folsome—Brittain experiment, the material of the microspheric membranes originates from formaldehyde. We may suppose that the synthesis of this membrane occurs by formose reaction. In this case, however, a stoichiometric interlocking must exist between the autocatalytic formose cycle and membrane formation, which is the basic stoichiometric condition of the existence of proliferating microspheres, as postulated theoretically previously.

The other basic condition for the division of proliferating microspheres is the two-dimensional liquid nature of the membrane surrounding the microspheres. There were no data about this in the paper of Folsome and Brittain. However, the statement of the authors that the microspheres may be blown up by osmotic shock effects refers to their two-dimensional liquid nature.

Thus, in the water droplets floating in the ascending convection air flow in the primordial atmosphere, both the conditions for the spontaneous formation of microspheres and the intermediates of the formose reaction were given. This means that if the membrane of microspheres is of a two-dimensional liquid nature, then in the water drops microscopic systems proliferating by division, (i.e., microspheres) must have appeared. Either the solution in the microspheres already contained the intermediates of the formose cycle, or these were formed later in the solution by formaldehyde diffusing into the sphere from outside.

Considering the whole atmosphere, this can be imagined so that these proliferating microspheres were formed and divided again and again in the billions of such water drops being destroyed by unfavorable conditions (i.e., freezing or evaporating), and then newly re-formed in the subsequently condensing water drops. Thus, according to the chemoton theory, systems proliferating by division were not only produced once, by chance as a result of some miracle, but rather countless times wherever the conditions for their formation were favorable. In fact, in the atmosphere of primordial Earth the conditions from a vast amount of formation and proliferation were provided everywhere.

Spontaneous genesis of chemotons

The appearance of proliferating systems is a very large step toward the genesis of life, but according to the criteria in Chapter VI, the proliferating microspheres as such cannot be regarded as living systems, even though they have a metabolism, they proliferate by division as bacteria do, and they satisfy some other criteria of life. What is missing, however, is the control by coded programs and, correspondingly, variability (which is the basis of biological evolution) is also lacking. As we have seen in the previous chapter, this role can be played by RNA in prebiotic chemotons.

Under the conditions of the primordial atmosphere, the formation of RNA precursors is possible, because ribose is produced in the formose reaction and, under UV irradiation, from N_2, NH_3, cyan and cyan derivatives appear from which, in turn, nucleotide bases are formed. If in the water drops the formation of RNA molecules can also take place, then in the enclosed liquid of the microspheres RNA may be present and thus chemotons may appear automatically in a single step, instead of, or besides, proliferating microspheres.

Whereas the formation of formose intermediates, RNA precursors, and microspheres necessarily must occur, as shown by chemical evolution experiments, the spontaneous formation of RNA molecules is a much less probable event. Nevertheless, it is by orders of magnitude more probable than the appearance either of macromolecules having a given sequence or of such specific RNA molecules as are assumed in the hypercycle theory.

Obviously, it is virtually impossible to calculate accurate probabilities with our present knowledge, but an approximate lower probability limit can be estimated, above which the formation of chemotons in one step should have a real meaning.

If the total water volume floating in the primordial atmosphere is supposed to have been 10^9 km^3, the volume of individual drops 1 mm^3, and we further assume that the average lifetime of a drop was 10^{-1} year and that the favorable conditions lasted for 10^6 years, then the number of drops formed during this period is 10^{34}. Thus if the probability for the spontaneous formation of an RNA molecule in the water drops of the primordial atmosphere exceeds 10^{-34}, the atmosphere should have contained drops with RNA molecules. Supposing that the volume of a droplet corresponds to that of 10^8 microspheres, then the probability for the formation of RNA molecules had to be higher than 10^{-26} in order to have microspheres enclosing RNA, i.e., that chemotons could form in a single step (when we suppose the formation of only one *de novo* microsphere in individual droplets).

The results of chemical evolution experiments suggest that the probability for the formation of RNA molecules is in fact orders of magnitude higher than that, especially if we consider that it is not the spontaneous formation of RNA molecules that is actually needed, because depending on temperature, oligonucleotides with 4–20 members may be enough to start template polymerization. Thus it seems that the one-step formation of chemotons had a real probability in the primordial atmosphere.

However, Slemmer (1982) showed that the *de novo* formation of chemotons in two steps, i.e., through microspheres, has a much higher probability. He assumed that in proliferating microspheres ribonucleotides and ribonucleosides are formed from the cyan, ammonia, and formaldehyde diffusing into the spheres (see Chapters VII and VIII), but in the proliferating microspheres they do not have any stoichiometric role. At the same time, the membrane of microspheres is impermeable for these products, thus the concentration of nucleosides and nucleotides increases continuously. With increasing concentration the probability for spontaneous polymerization also grows. Thus, inside the microspheres, the probability for the spontaneous formation of ribopolynucleotides with a template effect is much higher than in the extrachemotonial regions of floating drops. Upon consideration of the above it seems that not only the formation of proliferating microspheres, but also that of chemotons may well have been a process in the primordial atmosphere, as long as the necessary conditions for biogenesis existed.

External conditions of early evolution

The scientific literature uses the word "evolution" in two senses without, however, clearly distinguishing between the two meanings. One expression is mainly used for changes in the nonliving world (e.g., the evolution of the Universe, chemical evolution) but is sometimes also expanded to the living world (e.g., the evolution criterion of Prigogine). In this sense, the word usually means variations in a given system or in its parts showing a move in the direction of organization and complexity. The other sense is of biological nature: the elements of the system are here proliferating units and in this case, the variations occur in the sequences of generations of these proliferating units. Thus the variation here does not take place in the given elements of the system, but in their descendants, independently of whether the units of the system are capable of proliferation by themselves (e.g., living beings) or whether other parts of the system make them proliferate (e.g., viruses, technical goods). General theories of evolution have also been born in this second meaning of the word (Varela, Maturana, and Uribe, 1974; Csányi, 1978, 1980, 1981, 1982; Csányi and Kampis, 1984, 1985; Maturana and Varela, 1980; Gluchkov, Ivanov and Yanenko, 1983).

In Chapters VII and VIII we have used the word "evolution" in the first sense and will continue to use it so when speaking about chemical evolution. However, in interpreting prebiological and biological evolution, the second meaning will be used. Both are processes occurring through sequences of numerous generations of proliferating systems via eventual changes in individual descendants. Thus the basic condition for biological and prebiotic evolution is the possibility for the continuous and infinite proliferation of living individuals. Evolution in its biological meaning can exist only so far as the

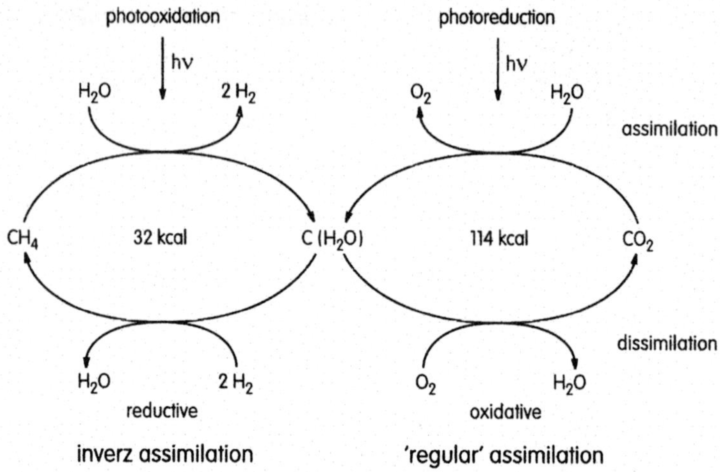

Fig. 9.1. According to Decker (1978), in contrast to the current carbon cycle including the partly and completely oxidized states, under reducing conditions an inverse cycle could take place involving partly and completely reduced forms of carbon. Energy is provided for both cycles by the Sun.

possibility for proliferation exists. When discussing the early evolution of chemotons, first we have to decide whether the conditions for infinite proliferation of prebiotic chemotons existed on primitive Earth.

According to our present knowledge, liquid water has been present continuously on Earth. However, nutrients – although they were present in abundance for the genesis of life – later transformed into substances not suitable as nutrients for primitive living systems, due to chemical evolution. For the prebiotic metabolic network shown in detail in the previous chapter, as an example, formaldehyde, cyan, simple cyan derivatives, and ammonia served as nutrients. However, these substances react quickly with each other thereby making the operation of chemotons possible

It should be emphasized again at this point that the prebiotic metabolic network shown is only a hypothetical variant, and probably numerous other operating networks may be created. However, concerning nutrients, there is not a great choice: only substances that have sufficiently small dimensions to diffuse through the membrane are suitable nutrients at the prebiotic level. Thus, in order to ensure the continuous proliferation needed for the evolution of chemoton-like prebiotic systems or other primitive cells, these simple organic and nitrogen compounds had to be continuously generated, i.e., carbon and nitrogen had to participate in a global cycle of which these simple compounds had to be members, also under the conditions of primitive Earth.

Decker (1978) showed that a cycle similar to the present carbon cycle might also operate under reductive conditions and that on primitive Earth this process (called by him "inverse assimilation"), could probably continuously provide the nutrients for the primitive living world (Fig. 9.1). In both processes, formaldehyde plays the central role (or its polymers, the sugars). Formaldehyde is an energy-rich compound in both reductive and oxidizing atmospheres as a result of its being partly a reduced, partly an oxidized carbon compound. It provides the driving force for the metabolism of living systems by its oxidation in an oxidizing environment, and by its reduction when in a reducing atmosphere. In both cases, it is regenerated by photo-energy, but whereas in an oxidizing atmosphere this is possible only by the specific photosynthesizing apparatus of living organisms, in a reductive atmosphere it is a spontaneous process.

The inverse assimilation hypothesis of Decker seems also to be supported by the fact that methanogen bacteria are the most ancient living beings known. Unfortunately, inverse assimilation by itself does not provide the possibility for continuous proliferation, because in this process only the parts of organic compounds participate, which then depart as by-products of the metabolism in the form of CH_4 or other gases transforming to CH_4. Thus the substances being incorporated into the living system, as well as the larger molecules formed by direct chemical evolution, continuously decrease the amount of organic substances participating in the cycle until they are totally consumed. The biological destruction of these larger molecules needs specific biochemical mechanisms (production of exoenzymes) that form only in a long evolutional process, not being present at the beginning of prebiotic evolution. If the genesis of life had taken place in the surface pools on primitive Earth or in the primordial ocean, then the organic substances should have been deposited in them as a precipitate of large molecules, as occurs in the current chemical evolution experiments.

However, if we accept the hypothesis of Woese, according to which life was generated in the floating water above the still hot surface of the Earth, the organic substances reaching the hot surface of the Earth (via the air currents) could be thermally cracked into smaller molecules that – in the reducing atmosphere – could then be reduced to methane or to other substances suitable as nutrients for primary living systems. This process may have provided continuously for the return of organic material (consumed by chemical evolution or by prebiotic living systems) into the carbon–nitrogen cycle and, through this, for the nutrient provision and thus continuous proliferation of prebiotic living systems. The period in which the surface of primitive Earth was sufficiently hot (above 1000 °C) is estimated to be several millions of years. This is obviously sufficient for evolution to develop heterotrophic nutrition ensuring thereby the cyclic transformation of organic materials on its own. The early stage of evolution, as we shall see later, might have been a very fast process.

The early evolution of chemotons

Equation [8.60] and Eqs. [8.61.1] – [8.61.19] that belong to it make concrete calculations concerning the amounts and concentrations of the components of prebiotic chemotons possible. Let us assume – based on experience from experiments on the formation of microspheres – that the diameter of prebiotic chemotons d=1 μm and that the area occupied by one membrane-forming molecule in the membrane,

$$S_T \approx 1\text{nm}^2 \qquad [9.1]$$

In this case, the number of T molecules on the total (internal and external) surface of the membrane of a prebiotic chemoton, S, is

$$m = \frac{S}{S_T} \approx 0{,}6 \cdot 10^7 \qquad [9.2]$$

Considering that, according to Eq. [4.65], the number of nucleotides in the RNA template molecule is identical to that of membrane-forming molecules (let us now exclude the terminal nucleotides):

$$\sum_{i=1}^{4} \kappa_i \approx 0{,}6 \cdot 10^7 \qquad [9.3]$$

If we divide this number by Avogadro's number, we obtain the number of moles of nucleotides used up for the replication of a template in one prebiotic chemoton (mole). Since the volume of a prebiotic chemoton is of such dimensions, $Q=10^{-15}$ dm^3, the nucleotide concentration in chemotons before triggering of template polymerization processes should exceed the concentration

$$c_{\sum_{i=1}^{4}\kappa_i} \approx 2\cdot 10^{-2}\,\text{mole}/\text{dm}^3 \qquad [9.4]$$

i.e., for prebiotic chemotons

$$V^* > 2\cdot 10^{-2}\,\text{mole}/\text{dm}^3 \qquad [9.5]$$

which seems to be a reasonable value.

The amounts of individual components can only be determined if the nucleotide composition, i.e., the values of κ_i, are known (NH_3, HCN). However, the amounts of individual nutrients also consumed [$(NC)_2$, HCN] depend on the nucleotide composition, i.e., on the actual κ_i values as determined by Eqs. [8.61.7] and [8.61.12]. This means that, for chemotons with different nucleotide compositions, different nutrient compositions provide the optimum conditions, and vice versa, under given environmental conditions the fitness of chemotons with different nucleotide compositions also differs. Thus from mutations occurring during the proliferation of prebiotic chemotons, those resulting in variation of the nucleotide composition are surely not neutral mutations, and therefore lead to a process that during generations may be regarded as evolution in the biological meaning, through continuous optimization (mutation + selection). At the same time, mutations changing the nucleotide sequence without varying the nucleotide composition are neutral ones (at least from the stoichiometric aspect), and thus do not represent an evolutionary force at this level of evolution.

If the primordial atmosphere above the hot Earth had a stable convection structure similar to Jupiter's today, the environment of prebiotic chemotons would have varied very much from place to place. In the lower layer the primary products of thermal cracking predominated, whereas in the upper layers of air those of photochemical processes. It should be emphasized again that the metabolic network of prebiotic chemotons shown in Chapter VIII is only an example. Numerous other network variants can be imagined, even more so as in this network no organic acids are involved although we know that they form in abundance during chemical evolutions, and that self-reproducing cycles can also be built from organic acids (see Chapter II). Thus if we consider only stoichiometric possibilities and limitations, we have to suppose that chemoton populations differing not only in nucleotide composition, but also in the type of metabolic network could have appeared, proliferated, competed, and evolved in the different layers of the primordial atmosphere.

However, fitness could already be influenced at that level not only by stoichiometric relations, but also by kinetic relationships within the limits of stoichiometric possibilities. Moreover, as one of the most decisive factors of evolution is the rate of proliferation, kinetic relations become the most significant factor of evolution for a given metabolic system. A living system capable of increasing hereditarily the rate of its reactions in a stoichiometrically optimized population has an advantage over others which cannot. In this process the sequence of nucleotides plays a basic role.

Appearance of Enzyme Activity

The problem of primitive catalysis

The most remarkable property of today's living beings, from the point of view of the biologist, is perhaps their enzymatic regulation. This ensures the fast reactions of the metabolism at normal temperatures, quick reactions to environmental changes, the manifestation of coded genetic properties, the realization of hereditary variations, etc. Thus it is quite obvious that biologists cannot imagine a living system without enzymes and that research concerning the genesis of life was first aimed at the search for the genesis of early enzymes or proenzymes, or that at least it supposed the existence of some specific primitive catalysts.

Nevertheless, after the clarification of the amino-acid sequence of enzymes it became clear that the probability for the spontaneous formation of enzymes with specific sequences is so small that biogenesis could not be imagined in this way. Thus, Bernal proposed, as early as 1959 that at the time of the genesis of life clay minerals could play the role of catalysts (Bernal, 1959a, b, 1967). Experiments aimed to clarify this problem have been reported and are in progress even today (Peacht-Horowitz et al., 1970, Peacht-Horowitz, 1974, 1983; Burton, Lohrmann, and Orgel, 1974; Lahav, 1975; White and Erickson, 1981; Hagan and Ferris, 1983; Coyne, 1983; White et al., 1983; etc.) and have been expanded to other mineralic materials (Gabel and Ponnamperuma, 1967; Reid and Orgel, 1967; Knobel et al., 1983; etc.), as well as to volcanic ash (Lavrentiev, Strigunkova, and Egarov, 1983). The catalytic effect of inorganic ions, especially metal ions, has also been studied (Löwenstein, 1958; Bahadur and Ranganayaki, 1983; Fakhral et al., 1981; Sawai, 1981; Ochiai, 1983). The heterocatalytic effect of inorganic precipitates has also been applied successfully, especially in the abiotic chemical processes of nucleotides (Burley, 1965; Lohrmann, 1982).

Among organic substances, the study of the catalytic effect of coenzymes and coenzyme-type compounds has been an obvious choice (Yount and Metzler, 1959; Yatko-Manzo et al., 1959; Kranpitz, 1969; Osterberg, Orgel, and Lohrmann, 1973). White (1983) observed the catalytic effect of oligopeptides in template polymerization, and Fox and co-workers studied the catalytic activity of thermal proteinoids and even that of nucleoproteinoid microparticles (Hardenbeck et al., 1968; Oshima, 1968; Rohlfing and Fox, 1969; Dose, 1974; Fox, 1973a, c, 1975; Nakashima and Fox, 1972). Finally, it should be mentioned that both Bahadur and Fox have written in many papers that Jeewanus and microspheres also possess various catalytic activities.

The majority of these experiments have been successful. Catalysis is not such a rare event under prebiotic conditions as has been thought. However, a catalyst, in order to be able to play a role in biological evolution, has to satisfy at least two further requirements in addition to its accelerating effect:

1. In the course of divisions of living systems, its amount has to be doubled, such that either it must be of a self-producing (self-reproducing) character, or the living system has to produce it in a regulated or controlled process.

2. It should have hereditary variability, or the information concerning it in the program producing the catalyst must be able to undergo hereditary variations.

Thus the question is not whether a prebiotic metabolism could be developed without enzymes. The results of different evolutional experiments showed that the presence of specific catalysts was not necessary for primitive metabolic processes. Buvet discussed in detail the possibility of nonenzymatic metabolism in several of his works (Buvet, 1974; Buvet and Le Port, 1973), and the metabolic network shown as an example in the previous chapter also consisted of nonenzymatic elementary reactions.

It was discussed in Chapter VI that the catalytic superregulation based on stoichiometric and kinetic regulations is necessary not for the living state in itself, but for survival in the present living world. If catalytic superregulation (i.e., enzymes) had not appeared in living systems, their evolution would have remained very restricted in comparison to present possibilities.

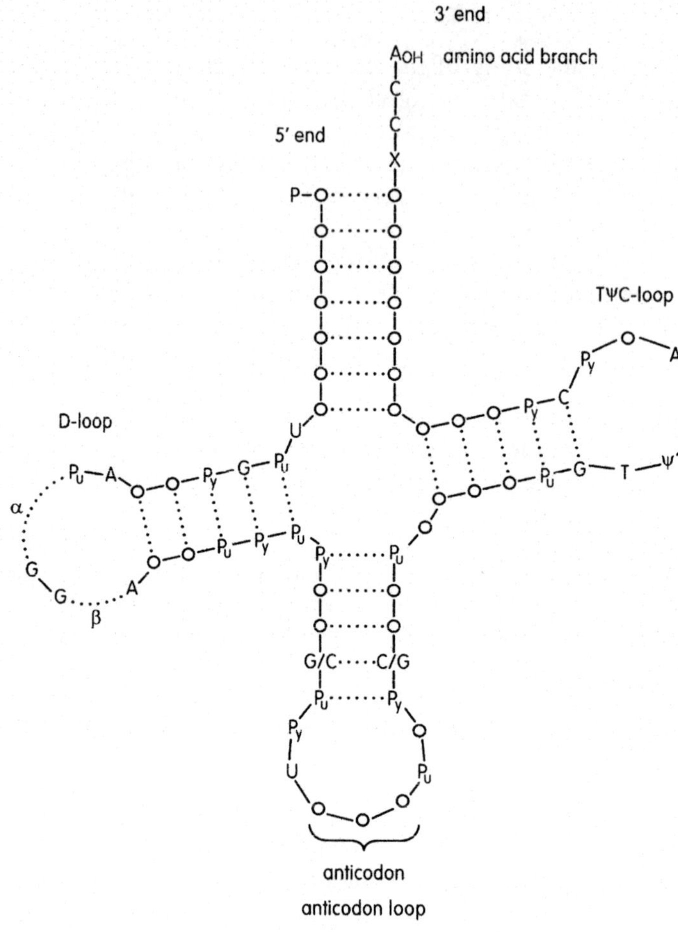

Fig. 9.2. The planar structure of tRNA molecules, the so-called cloverleaf structure.

Thus if we want to deduce the development of today's living world from primitive systems, we have to answer the following question: how did catalysts satisfying the former two criteria (i.e., the enzymes) develop? The inorganic catalysts of the external world can satisfy neither of the two requirements. Simple organic compounds (e.g., coenzymes) may satisfy the first criterion, and this requirement is also surely satisfied by Jeewanus and the membranes of microspheres, if their proliferating nature can be proved. However, the second criterion is not satisfied by either of these possibilities.

The conclusion that the appearance and development of enzyme activity in the course of prebiotic evolution took place in connection with RNA rather than with proteins was drawn by Crick (1968) based on his studies concerning genetic codes by Brewin (1972) on studying the polymerase function, by White (1976) on investigating the origin of coenzymes, by Gánti (1979c) on the prebiotic application of the chemoton theory, and by Korányi (1982b) on the analysis of the general spatial structure of tRNA. Altman concluded, from a quite different aspect – the study of the ribonuclease P enzyme – that for the activity of this enzyme its protein component is probably not responsible, but rather the RNA part occurring together with it (Lewin, 1984). Brewin (1972) assumes, from structural considerations, that in the primitive metabolism ribonucleic acids may have played the role of nucleic acid polymerase enzymes. However, earlier Crick (1966) interpreted tRNAs as Nature's experiment to let RNAs play the part of proteins.

Nevertheless, these assumptions have either not been taken into consideration at all, or have been qualified as absurd, as no example has been found illustrating the existence of enzyme RNAs (eRNA). Still, numerous known facts favor their existence, as the different RNAs have different tertiary structures determined exactly by their sequence, similarly to enzyme proteins. In this structure helical and nonhelical (looplike) parts alternate, as in enzyme proteins. The nonhelical parts contain reactive groups pointing to the outside, which may serve in enzyme reactions as active sites. Harvey and McCammon (1981) showed, by the investigation of the $tRNA^{Phe}$ of yeast, that the tertiary structure has flexible parts. It is obvious that the presence of such flexible parts is very favorable from the point of view of supposed enzyme activity. The structure and function of tRNAs have been known for two decades. This function, though it cannot be regarded as a real enzyme function, is very similar to it because in this case also a genospecific macromolecule of a spatial structure determined by a sequence performs substrate-specific manipulation on different molecules with the aid of specific recognition sites. The two-dimensional structure of tRNA molecules ("cloverleaf structure") is shown in Fig. 9.2.

Attitudes have changed significantly since that time; Cech and co-workers reported in 1983 that the r-RNA precursor of *Tetrahymena* catalyzes its own chemical transformation in a way that it cuts out a section from itself and binds it into a cycle (Zang et al., 1983). The operation performed by this RNA on itself is a typically enzymatic operation (cleavage and formation of covalent bonds), although the RNA in this case cannot be regarded as an enzyme because it does not remain unchanged at the end of the reaction, namely it performs a transformation upon itself. In the same year, Altman and co-workers proved that, in *E. coli* and *B. subtilis*, the RNA component of the macromolecule is responsible for the ribonuclease P activity, i.e., they discovered the first real enzyme RNA (Guerrier-Takada et al., 1983); and in 1985 they described the eRNA of *Salmonella typhimurium* and the gene sequence belonging to it (Baer and Altman, 1985). In 1984 they prepared *in vitro* the RNA component of the ribonuclease P gene by transcription and proved that its activity is identical to that prepared from *E. coli* (Guerrier-Takada and Altman, 1984). This is proof that eRNAs not only may have existed, but that they also exist today.

Evolution of RNAs

Let us now deduce the evolutional development of specific eRNAs from the nonenzymatic prebiotic chemotons by means of the chemoton theory (Gánti, 1979c). It is expedient to start from the general equation of chemotons, which is given for the simplest abstract chemoton model by Eq. [4.67]. By the application of this equation for chemotons containing a k-membered cycle and using X_i nutrients we obtain:

$$\left[m\sum_{i=1}^{k}\alpha_i A_i + gn\,pV_n + \boxed{T_m} \right] + m\sum_{i=1}^{1} X_i \xrightarrow{\;①\;}$$

$$2\left[m\sum_{i=1}^{k}\alpha_i A_i + gn\,pV_n + \boxed{T_m} \right] + m\sum_{i=1}^{1} Y_{i'} \qquad [9.6]$$

where gn = m and n is the number of nucleotides, in one template, and g the number of templates in one chemoton.

Using the analogy of Eq. [3.33], let introduce designations

$$\sum_{i=1}^{k}\alpha_i A_i \equiv \overline{A_i} \qquad [9.7]$$

$$\sum_{i=1}^{1} X_i \equiv \overline{X} \qquad [9.8]$$

$$\sum_{i=1}^{1} Y_i \equiv \overline{Y} \qquad [9.9]$$

and let us further consider that, under prebiotic conditions, instead of homopolymer pV_ns, RNA molecules built of the double helix of RNA containing four types of nucleotides were present. For this case, Eq. [9.6] has the following form if we take Eqs. [9.7], [9.8], and [9.9] into account:

$$\lfloor m\overline{A} + gRNS_n + \boxed{T_m} \quad] + m\overline{X} \xrightarrow{\;①\;}$$
$$\lfloor m\overline{A} + gRNS_n + \boxed{T_m} \;] + m\overline{Y} \qquad [9.10]$$

First, the number of template molecules in one chemoton should be determined. From Eqs. [9.2] and [4.65], in prebiotic chemotons

$$g \cdot n \approx 10^7 \qquad [9.11]$$

i.e., the total number of nucleotides in the templates of a chemoton is about 10^7. In order to determine the number of template molecules, their length (i.e., the polymerization degree of RNA) should be known. When providing the stoichiometric description of the metabolic network for prebiotic chemotons, we assumed, in the calculations of Chapter VIII, that RNA is a single long molecule consisting of two strands having only two terminal nucleotides (this explains the -2 member in the multiplication factor of R in Eq. [8.32]). This assumption has been made only to simplify calculations, as RNA molecules of such dimensions could not have existed under prebiotic conditions and do not exist even now. According to several authors (Kuhn, 1972; Eigen and Schuster, 1979; Gánti, 1979c), under prebiotic conditions, the assumption of polyribonucleotides with 10^2 nucleotides seems to be reasonable. This size corresponds in its order of magnitude to the size of today's functional RNAs; e.g., tRNAs contain 0.8×10^2 and eRNAs (ribonuclease P RNAs) 3–4 x 10^2 nucleotides:

If we accept that

$$n \approx 10^2 \qquad [9.12]$$

we obtain

$$g \approx 10^5 \qquad [9.13]$$

Thus we may calculate that in a chemoton with a diameter of 1 μm there are about 10^5 double-helix RNA molecules consisting of 100 nucleotides, on the average. This RNA stock is distributed statistically between the two descendant chemotons on the division of the parent chemoton. As we have seen, concerning their overall RNA composition chemotons adapt themselves evolutionally to their environment. In this phase, the nucleotide sequence of RNA molecules is completely random and different for each RNA molecule. Sequences have no functional role at all. However, if we disregard replication defects, the sequences are replicated and thus they are essentially inherited from generation to generation. On the other hand, the probability of replication defects is by orders of magnitude higher under prebiotic conditions that in today's living beings, not only because the various reparation mechanisms are missing, but mainly because replication is realized here by nonenzymatic processes. Both Eigen and Kuhn calculate that the probability of replication defects under prebiotic conditions is $W = 10^{-2}$, where W is the probability that, on the newly built strand, a randomly chosen nucleotide basis does not correspond to its counterpart in the template (Eigen, 1971; Kuhn, 1976). We also accept this value, but with the restriction that the modification of RNA molecules is not only the result of point mutations.

However, with such a high mutation probability, hardly any exact replication can occur, and about 10^5 bases of the descendant chemotons differ from the templates of the parent RNA. Thus variability is extremely large, and if this is accompanied by an appropriate selection force, a very quick evolution may result. On the other hand, for point mutations the environmental nutrient composition does not constitute a selection force capable of causing competition between the sequences. But if among these sequences one or the other can catalyze some of the elementary steps of the metabolic network of prebiotic chemotons, the metabolism of these chemotons will somewhat accelerate relative to that of the others, their generation time will be shortened, and thus, because of their increased fitness they will have an evolutional advantage over the other chemotons. This was unambiguously proven by computer simulation (Békés et al., 1980).

The probability that RNA sequences specifically catalyzing some of the elementary reaction steps appear as a consequence of point mutations is very small. Thus it is expedient to examine whether there are molecular mechanisms operating in prebiotic RNA replication that necessarily lead to the *de novo* appearance of enzyme RNAs.

Development of enzyme RNAs

In addition to hereditary changes resulting from point mutations, variations may also take place in RNA molecules that lead to the development of globular RNA molecules, and this variation is also hereditarily determined. One of these types of mutations is described by Kuhn (1972), and another mechanism is provided by Gánti (1979c, 1983). Both mechanisms have in common that they are based on the combination of RNA chains.

Kuhn's mechanism may become operative when two double-stranded RNAs having an identical sequence interact in such a way that the 5' end of the (+) strand in the first molecule pairs up with the 3' end of the (minus) strand in the other one, and, simultaneously, the 3' end of the (minus) strand in the first molecule binds the 5' end of the (+) strand in the second molecule. If these strands are covalently bonded through 3'–5' bonds, a sequence is obtained in which the beginning and the end of the strand have a complementary structure, thus being capable of turning into an intramolecular double helix, i.e., of loop formation (Kuhn, 1972). The process is shown in Fig. 9.3. As the number of possible sequences in a 100-nucleotide RNA is 10^{60} and as there are about 10^{5} RNA molecules in a typical prebiotic chemoton, loop formation via this

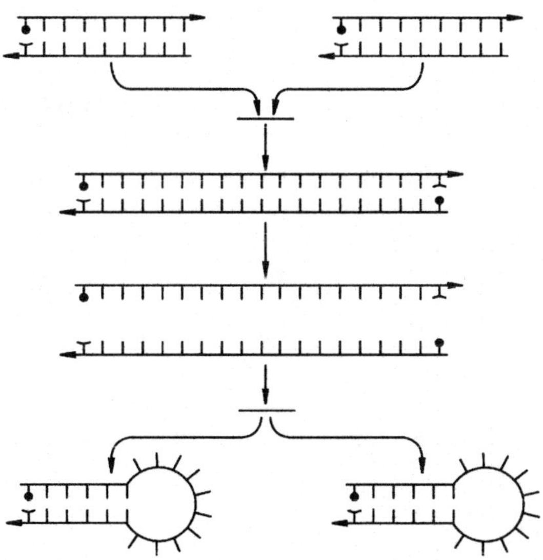

Fig. 9.3. A possible mechanism of the formation of intramolecular loops (Kuhn, 1972). This mechanism presumes the chemical coupling of two RNA double strands of identical sequences.

mechanism is practically impossible, as there is no finite probability that in the same chemoton, RNA molecules of the same sequence could meet.

The other mechanism (Gánti, 1979c) describes the formation of complementary sequences and thus that of loops as a natural, alternative reaction possibility of template polymerization in the course of prebiotic, nonenzymatic replication. When the formation of a new strand on the RNA template is sufficiently advanced, then either a new nucleotide is bound to the new strand (corresponding to the sequence of the template) thereby increasing the length of the strand, or the end of the template strand folds back and gets bound there, thus forming a loop. The probability of this latter event is especially high if a sequence section is present in the chain that is complementary to the sequence at the end of the same strand. This is in fact a frequent phenomenon. The mechanism is shown in Fig. 9.4.

Thus looplike RNA structures must appear. We may therefore assume that the major part of the RNAs in prebiotic chemotons is likely to have been of a palindromic sequence, i.e., of a loop structure. These loop-structured RNAs could probably also react with each other, thereby developing larger, globular RNA molecules containing more than one loop (Kuhn, 1972; Gánti, 1979c). As an example, the prebiotic formation of an RNA molecule similar to the cloverleaf structure of tRNAs is shown according to this mechanism in Fig. 9.5 (Gánti, 1979c).

In this case, spatial structure is already determined solely by the sequence of nucleotides, i.e., in this phase the spatial structure of RNA molecules is also inherited. On the loop sections, i.e., on non-pairing sections, a great variety of atom groupings appear that are suitable for the formation of different bonds: unpaired electrons as well as positive or negative excess charges, together with groups capable of hydrogen bondings, because in these regions the amino, imino, and oxo groups of the nucleotide bases are free, and the phosphate groups also have free =O or –OH groups. Loops especially have many possibilities for forming low-energy bonds, such as hydrogen bonds, which may be cleaved even at ambient temperatures by thermal motion. The metabolic intermediates of prebiotic chemotons as potential substrates may thus interact with the RNA loops present. These bonds are formed and cleaved over and over again.

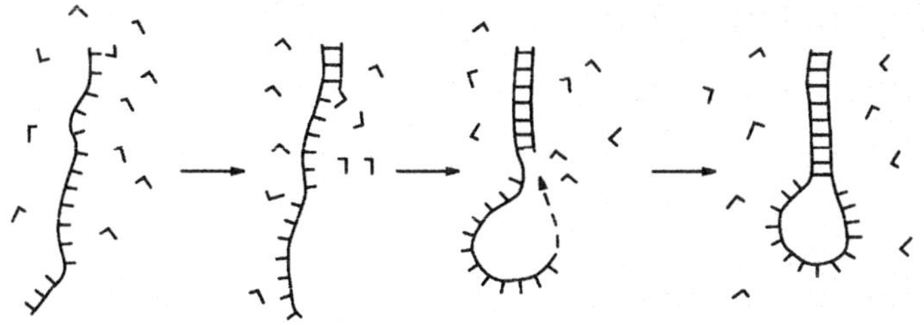

Fig. 9.4. A much more probable mechanism of the formation of prebiotic RNA loops: instead of a free nucleotide, the free end of its own strand moving via thermal motion is bound to the new strand synthesized on the RNA template. Thus a single-stranded loop is formed in which the two ends are of complementary sequences (Gánti, 1979c, 1983).

A predetermined spatial structure, functional groups capable of bond formation, and the possibility for temporary interactions are actually the preconditions of enzyme activity. The only question is whether all of these could have developed *in the configurations corresponding to the given substrates.* In other words, among the RNA molecules formed, could there have been present such that were capable of catalyzing one or another step of the metabolism of prebiotic chemotons?

Temporary bonds between potential substrates and the chemically active groups of loops can be established not only when the globular RNA is already formed, but also preceding this phase, i.e., when loop RNAs constituting the globular RNA are still in their free state. However, the temporary bonds forming between the potential substrates and the free loop RNAs themselves can direct the combination of such loop RNAs that make possible the formation of globular RNAs capable of performing enzymatic reactions on the substrate (Gánti, 1983). This process is illustrated schematically for cleaving enzymes in Fig. 9.6, and for synthesizing enzymes in Fig. 9.7. It is apparent that the potential substrates themselves construct the enzymes needed for their own transformation. This mechanism shows that the appearance of macromolecules having enzymatic activity is not a consequence of a random event, but is rather the result of a process that under prebiotic conditions must lead to the appearance of template polymerization and chemotons, and inside the chemotons to that of RNA loops, then to the construction of eRNAs from the loop RNAs. According to our computer simulations, the more metabolic steps are catalytically accelerated in a chemoton, the shorter its generation time, i.e., the greater its fitness under prebiotic conditions. This represents a very strong evolutional force for the development of chemotons, all the metabolic steps of which are catalyzed enzymatically (in this phase by eRNAs).

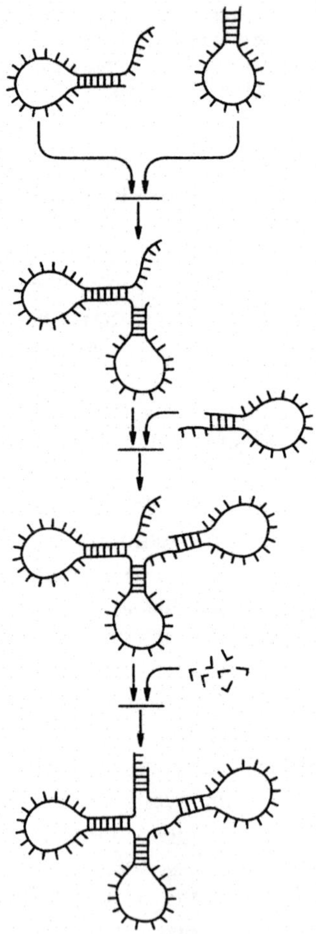

Fig. 9.5. With the coupling of arbitrary loop structures, globular RNAs may be formed having a sequence-determined tertiary structure. Here the prebiotic formation of a cloverleaf structure is shown (Gánti, 1979c).

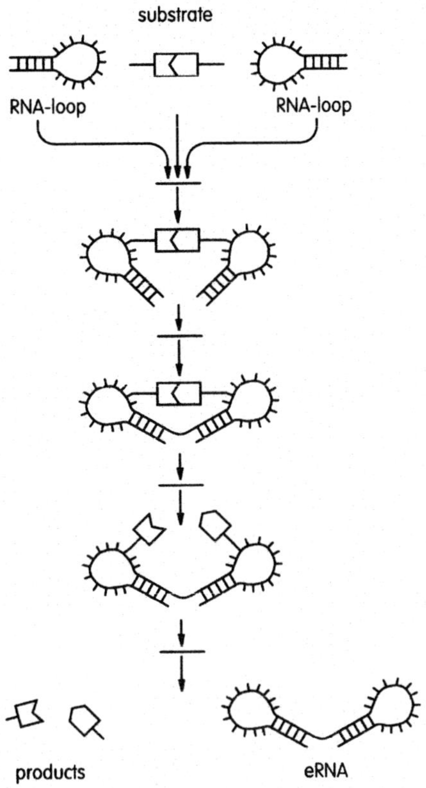

Fig. 9.6. The assembly of a cleaving enzyme RNA.

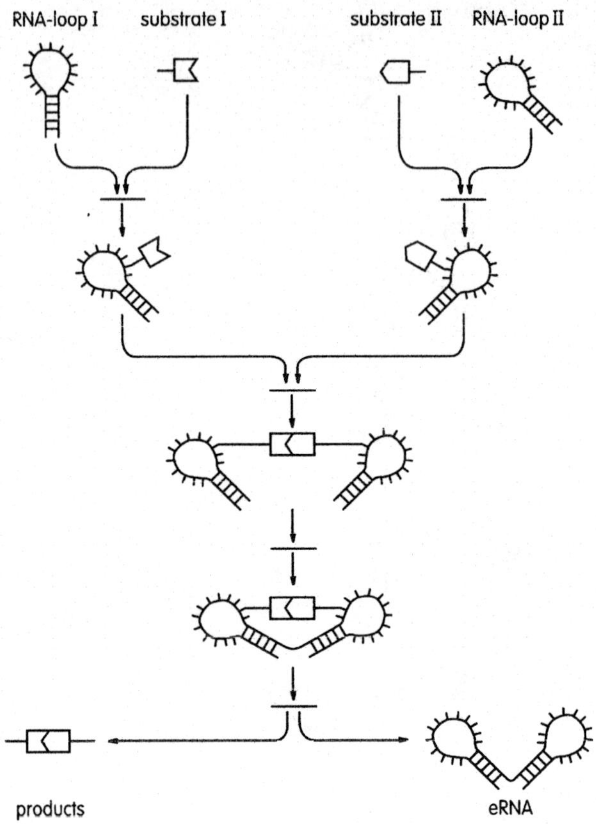

Fig. 9.7. The scheme of enzyme assembling for synthesizing enzyme RNA.

The Origin of Coenzymes

General considerations

By "coenzyme", we usually mean the non-protein part of protein enzymes, dissociating from the protein part but indispensable for the enzyme activity. In a broader sense, however, the nondissociable but nonproteinic parts of enzymes indispensable for enzyme activity (prosthetic groups) are also considered to be coenzymes. In what follows, the coenzyme concept will be used in this latter, broader sense. More than half of the known enzymes require coenzymes for their operation (IUB, 1979; White, 1982). Enzymes undergo a continuous and well-recognizable transformation in the course of biological evolution. On the other hand, coenzymes, the majority of which consist of complex organic molecules, have not changed essentially during this evolutional process. Thus the conclusion can be drawn that the evolution of coenzymes was complete before the beginning of the diversification of enzymes, i.e., before the appearance of the genetic code and the translation apparatus (Reanney, 1977; White, 1982).

It has already been mentioned that some coenzymes can catalyze their characteristic reaction even in the absence of the apoenzyme. Thus, with respect to thiamine it has been shown to possess decarboxylase activity (Mizuhara and Handler, 1954; Yount and Metzler, 1959; Yatko-Manzo et al., 1959; Kranpritz, 1969) and ATP transphosphorylase activity (Lőwenstein, 1958); with respect to adenosine cobalamine, under physiological conditions it can participate in mutase-type reactions in the absence of enzymes (Dowd and Shapiro, 1976); and with respect to pyridoxal phosphate, it can catalyze all four types of reactions in which it has an enzymatic role (transamination, racemization, decarboxylation, and elimination of β-hydrogen), also in the absence of the enzyme (Metzler, Ikova, and Snell, 1954). A similar phenomenon has been observed with flavins in photooxidation and dehydrogenation reactions (Haas and Hemmerich, 1979).

Based on the above, the early assumption could be made that, preceding the appearance of enzymes, the coenzymes played the role of catalysts in the primordial metabolism (Eakin, 1963). However, the spontaneous and primary formation of coenzymes under prebiotic conditions is, in the majority of cases, questionable. In the case of nucleotide coenzymes (as was discussed in detail in Chapter VII), though bases and ribose are readily found, the prebiotic formation of di- and triphosphates seems quite improbable. In dinucleotide coenzymes the unusual 5'–5' diphosphate bonding is found. The possibilities for its formation have not been studied experimentally although, as seen in Chapter VIII, Hidvégi has assumed that such intermediates may be formed during the synthesis of nucleotide imidazolides. Among the groups participating in the construction of CoA (ADP, pantothenic acid, β-alanine, cysteamine) only the formation of β-alanine is proved (Miller, 1955; Miller and Urey, 1959); its prebiotic occurrence is thus not probable.

Vitamin B_{12} has an outstanding role in prokaryotes known as the most ancient cells (Georgopapadokou and Scott, 1977). Among its components, nucleosides and imidazole or methylimidazole appear directly in the prebiotic chemical reactions. There are early references concerning the fact that porphyrinlike compounds are among the products of prebiotic chemical reactions (Hodgson and Baker, 1967; Hodgson and Ponnamperuma, 1968; Hayatsu el al., 1972). However, their nature has not as yet been investigated and, on the other hand, there are no recent data in the literature confirming the earlier ones. It seems probable that the raw materials needed for the formation of porphyrinic compounds may indeed be formed under prebiotic conditions, but the spontaneous formation of coenzymes is doubtful.

There is no direct experimental proof for the prebiotic formation of thiamine pyrophosphate, pterins, and pyridoxal phosphate. Upon summarizing the above results, it seems quite improbable that, in the prebiotic metabolism, the role of catalysts may only have been played by spontaneously formed coenzymes.

Coenzymes are the remnants of eRNAs

Harold H. White (1976, 1982) was the first to point out that the majority of coenzymes are nucleotides and that these nucleotide coenzymes may be the remnants ("fossils") of the active groups of primitive eRNAs. On conducting further studies, Korányi also established that a great part of non-nucleotide coenzymes are related to nucleotides and thus White's hypothesis can also be extended to include them (Korányi and Gánti, 1981; Korányi, 1982a,b).

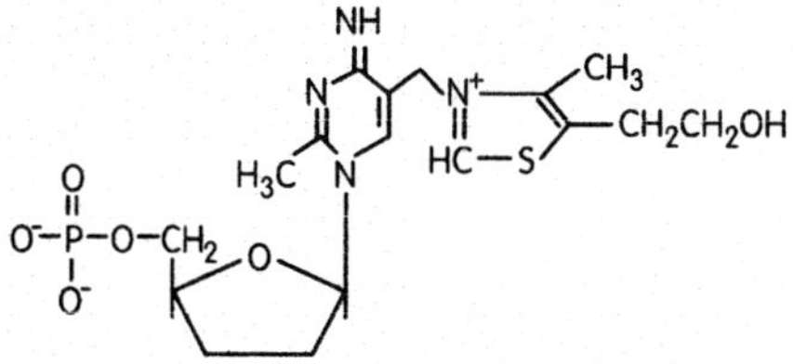

Fig. 9.8. According to White (1976), in prebiotic metabolism, not thiamine pyrophosphate, but one of its precursors, thiamine mononucleotide, itself a nucleotide, participated.

This relationship is trivial for mononucleotide coenzymes. It is also obvious for dinucleotide coenzymes, though the 5'–5' phosphodiesteric bond occurring in them is absent in today's nucleic acids. On the other hand, it was shown in Chapter VIII that the formation of such intermediates may be assumed in the synthesis of nucleotide imidazolides. Flavin nucleotides are built of a nucleotide and a pteridine derivative, a cyclic, N-containing isoalloxazine. However, the synthesis of pteridines starts from guanosine triphosphate; coenzymes with a pterin framework (FAD, FMN, folate) are thus also related to nucleotides. Thiamine pyrophosphate is a pyrimidine derivative condensed with a thiazolium derivative. According to White (1976), in the prebiotic metabolism, not thiamine pyrophosphate but rather one of its precursors, thiamine mononucleotide, itself a nucleotide, took part. The compound is shown in Fig. 9.8. Though pyridoxal phosphate is not a nucleotide, its ring part is all pyrimidine derivatives, and thus it is also related to nucleotide bases.

CoA is built of a complete ADP and pantotheine bonded to its terminal phosphate. In the B_{12} coenzyme there are even two nucleotides, a dimethylbenzimidazole nucleotide and a 5'-deoxy-5'-adenosyl group. Dimethylbenzimidazole is surprisingly similar to the ring configuration of the isoalloxazine in flavins (Fig. 9.9); the benzimidazole ring of the B_{12} coenzyme even originates from one of the precursors of riboflavin, and thus its

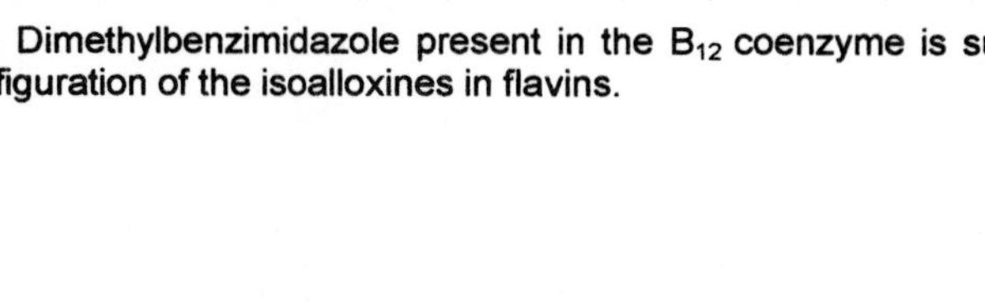

Fig. 9.9. Dimethylbenzimidazole present in the B_{12} coenzyme is surprisingly similar to the ring configuration of the isoalloxines in flavins.

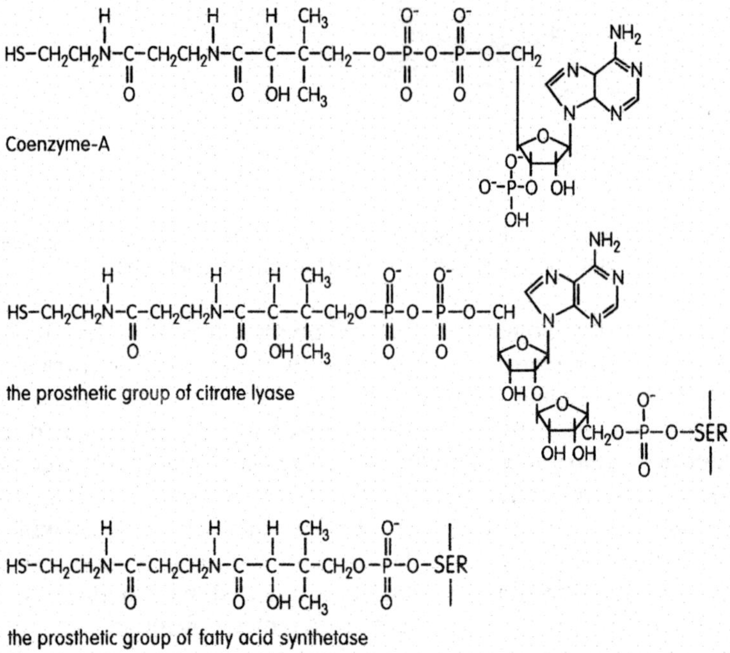

Fig. 9.10. Derivation of 4-phosphopantotheine from nucleotide-containing compounds (White, 1982).

biosynthesis also starts from a nucleotide. Even prosthetic groups having no ring parts, such as 4-phosphopantotheine (needed for the operation of citrate lyase or fatty acid synthetase) can be derived from nucleotide-containing compounds (White, 1982). The assumed evolutional steps are shown in Fig.9.10.

It is worth mentioning that the active groups of several enzymes operating without a coenzyme can also be derived from nucleotides (White, 1976; Korányi and Gánti, 1981). In the active sites of aldolases catalyzing one group of lyase reactions, histidine is always present, the imidazole nitrogens of which play a decisive role – together with the lysyl-NH_2 group – in catalysis. The imidazole in histidine is closely related to the rings of nucleotide bases. Moreover, histidine is the only amino acid the biosynthesis of which starts from a nucleotide (ATP) and not from metabolic intermediates. In addition to aldolases, nucleases, some phosphotransferases, dehydrogenases, phospholipases, and DNases also contain histidine in their active centers.

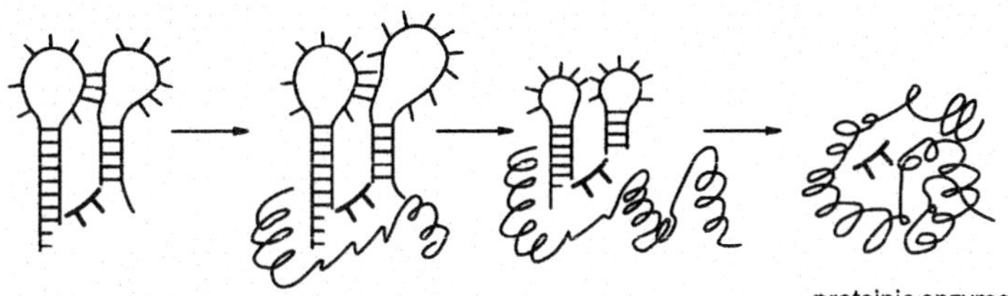

Fig. 9.11. Accordings to White, the parts of eRNAs having mostly structural functions were gradually substituted by protein after the appearance of translation.

White's opinion was that the active sites of one-time globular eRNAs were left behind as coenzymes, whereas the parts of eRNA having structural functions were replaced by proteins after the appearance of translation (Fig.9.11). The role of histidine as an imidazole derivative in the active center of protein enzymes is especially interesting. In today's RNA, imidazole plays no role, but as seen in Chapter VIII, primitive RNAs might have been synthesized from imidazolides; thus at the 5' end of each eRNA molecule an imidazole part had to be present, which could have an important role in the catalytic activity of eRNA. This function may well have been preserved in today's proteins through histidine.

Naturally, it is not necessary to assume that every current coenzyme was part of the structure of primitive eRNAs. A part of them could have played the role of a coenzyme even then, as the nucleotide part of the coenzyme might have been temporarily bonded to the RNA by hydrogen bonds for the duration of the reaction.

Nevertheless, it is a fact that not every coenzyme's origins can be traced back to nucleotides. Such exceptions include biotin and lipoic acid or, when taking the coenzyme concept in its broader sense, chinones, cytochromes, heme, etc. It cannot be either that, during the evolution of proteins, new coenzymes have been formed in connection with the evolution of metabolic networks. Moreover, it is certain that all of the enzymes operating in oxidizing atmospheres have entered metabolism not earlier than 2.5–3.0 billion years ago. Thus it is not surprising that in those oxidoreductases that can be regarded unambiguously as primitive enzymes 85.4% of the coenzymes are of nucleotide character, whereas in those developed after the atmosphere had become oxidizing, this proportion is only 14.7% (Korányi, 1982a,b).

The Origin of Genes and the Code

The origin of genes

It was discussed in detail in Chapter II that, if the template is a copolymer, i.e., it is built of different monomers, two types of reading mechanisms may be conceived pairing either between identical molecules (homologous pairing) or between different molecules (complementary pairing). In the former case the new strand built on the template is of the same sequence as the template, in the latter case it is of a complementary nature (negative copy). The copy of the negative strand is, however, also positive in the latter case, thus every second generation of the copies is identical.

As is known, RNAs participate in complementary pairing, i.e., the RNA strands in the eRNA chemotons are of a complementary character in half-half proportion. Based on Eq. [9.10] this case may be described as follows:

$$\left[m\overline{A} + \frac{g}{2}RNS_n^+ + \frac{g}{2}RNS_n^- + \boxed{T_m} \right] + m\overline{X} \xrightarrow{\;①\;}$$

$$2\left[m\overline{A} + \frac{g}{2}RNS_n^+ + \frac{g}{2}RNS_n^- + \boxed{T_m} \right] + m\overline{Y} \qquad [9.14]$$

It was also discussed in Chapter II that, for template polymers, the double strand is the stable form. If the RNA molecules are not of a palindromic sequence, i.e., the strands are not capable of forming a helical structure by themselves, then Eq. [9.14] has no meaning; in this case positive and negative strands are pairing with each other and form double-stranded molecules, and the state of the system in this case is described by Eq. [9.10]. However, if the RNA strands can build a double-stranded structure by themselves, i.e., they are capable of intramolecular pairing – and this is the case with eRNAs – then two kinds of stable helical forms are possible: the filamental double strand formed from the (+) and (–) strands, or the helical structure built separately from either the (+) or the (–) strand, these latter being mainly of a globular nature.

In the course of template polymerization, primarily double-stranded RNA molecules are formed independently of whether their sequence is palindromic or not. However, the ends of the double-stranded structure continuously open up and close as a result of the breathing mechanism discussed in Chapter II. If the sequence of RNA is palindromic, a closing up that follows an opening may have two different outcomes depending on whether the double-stranded structure is rodlike, i.e., stiff, or elastic to the extent that the two ends of the double strand can meet. In the first case, openings are always followed by closing up with the partner strand, consequently the structure remains unchanged (Fig. 9.12a), whereas opening up in the elastic case may be followed by a closing with the strand's own opposite end (Fig. 9.12b). In this latter case, however, the structure consisting of two separate strands opens in a zipperlike manner, the strand paired with its own and becomes helical intramolecularly, and thus the two strands separate from each other. Subsequently the other strand also becomes helical via an intramolecular mechanism.

This latter process is irreversible and must take place for appropriately elastic structures. Thus in chemotons containing eRNAs, the helical structure of the eRNAs is destroyed on reaching the concentration exceeding the critical value (V^*) of nucleotides (at the same time, their enzymatic activity ceases to exist). Replication then starts on the individual strands thereby forming double-stranded, helical RNA filaments, then their strands separate and take a globular, tertiary structure corresponding to their sequence by intramolecular pairing (at the same time eRNAs become enzymatically active again).

The sequences of the two strands are naturally complementary rather than identical. From this it follows that where a palindromic sequence is present in the one strand, a palindromic sequence will also be found in the other strand, i.e., the two strands – starting from opposite directions – contain parts of identical lengths capable of intramolecular helicalization, and the two strands have – from opposite directions – identical secondary structures, having at the same time different primary structures. Thus their tertiary structure is probably – but not necessarily – similar or identical. However, since the functional groups of the structurally identical sites are different even in identical tertiary conformations because of the different sequence, the replicas of eRNA molecules have no enzyme activity.

Let us denote, in what follows, the RNA strand of the complementary strands having enzyme activity by (+) and the one not having it by (−). It is obvious that eRNAs are not directly self-reproducing because the negative strand, not eRNA, is synthesized on them. Neither are the negative strands self-reproducing, because eRNA are produced on them, i.e., positive strands. Thus the negative strands carry the information related to eRNAs coded in their sequence, though they do not possess enzyme activity. If the sequence of the (minus) strand is changed, the sequence of eRNA coded in it also changes. With this the property of eRNA is also changed, and this change is hereditary. Thus the negative strand is the gene of the eRNA. This means that during prebiotic evolution not only the role of enzymes, but also that of genes was played by ribonucleic acids. Further, not only the formation of enzymes, but also that of genes was an essential process in the evolutional chain of events.

The origin of RNA polymerases

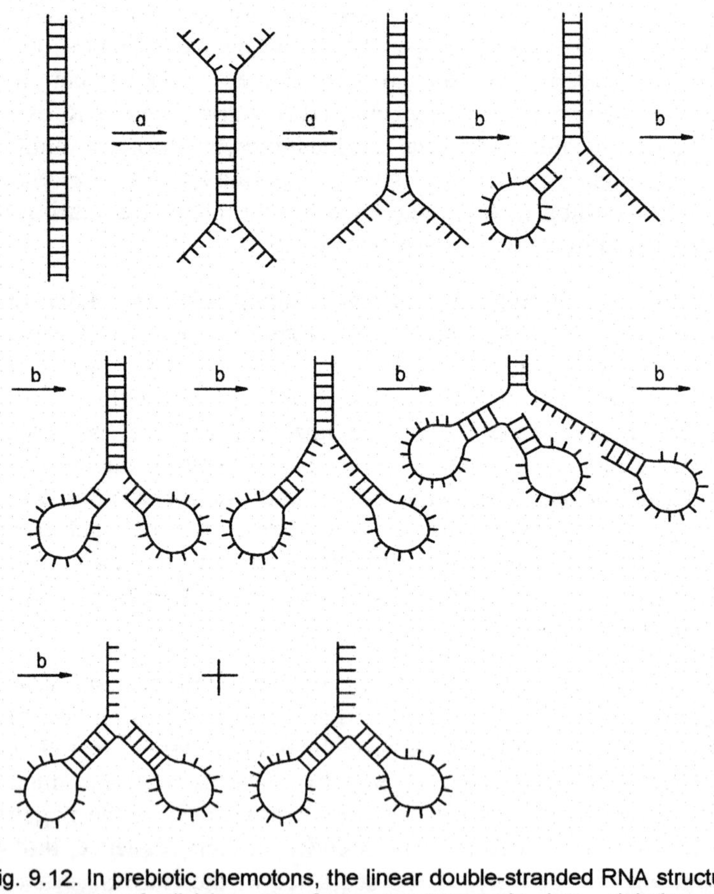

Fig. 9.12. In prebiotic chemotons, the linear double-stranded RNA structures synthesized in the course of division transform spontaneously into globular structures due to transformations in direction b.

In this phase, fitness is determined only by the rate of proliferation, i.e., by the length of generation time. This, in turn, depends partly on the harmony of the metabolic network and available nutrients, and partly on the reaction rates of the elementary reaction steps. Let us now investigate, in the case of a metabolic network stoichiometrically optimized to its environment, what changes may lead to a decrease in generation time (i.e., to an increased fitness) in the descendants of a chemoton in which the first eRNA molecule has appeared.

1. The number of molecules of a given enzyme (i.e., the concentration of the enzyme) increases in descendants preceding division, a replica of the enzyme is made, thus in the dividing chemoton two eRNA molecules are already present. Because RNA molecules are distributed statistically between the two descendants, the probability that the daughter cells contain 1–1 eRNA molecule is 50%, and the probability that one or the other of the daughter cells contains both eRNA molecules is 25–25%. Thus, based only on statistical distribution, there is a 50% probability that one of the descendants will be richer in the enzyme than the parent chemoton, thus its generation time will be shorter and consequently its fitness will be higher than that of either its companion or parent. It is obvious that in this way the concentration of a given enzyme is very quickly optimized in the evolutional process.

2. A more efficient enzyme of the given reaction appears in the descendants. This is the result either of a mutation of the preformed enzyme or of a *de novo* enzyme formation mechanism (described earlier). Because the probability of mutation defects is very high ($W \approx 10^{-2}$), the hereditary variability of enzymes is also very large. Obviously only a small part of these varieties have higher enzyme activity than the original molecule. However, based on the large variability, the optimization of enzyme activity is an evolutionally fast process at this level, though it is somewhat lower than the optimization of enzyme concentrations.

3. In addition to the existing enzyme, another one catalyzing another reaction also appears. This is possible according to the *de novo* enzyme formation mechanism, based on the presence of metabolic intermediates as potential substrates and also on the high mutation probability, as long as all the appropriate enzymes needed for the elementary steps of metabolism are developed. The appearance of each enzyme is naturally followed by its qualitative optimization on the one hand, and on the other hand, the concentrations of the newly formed enzymes have to be optimized again. Optimum concentration is – in the presence of more than one enzyme – determined by the maximization of the rate of the whole

network rather than by that of individual reaction steps. Considering that, according to the hypothetical mechanism, the substrates themselves "assemble" their own enzymes; we may assume that, on the evolutional time scale, prebiotic systems reach very quickly the state whereby every elementary step of the metabolic network capable of being catalyzed at all is regulated by eRNA.

Since all three evolutional processes seem to be fast, it is probable that at the beginning of prebiotic evolution, chemotons could reach in a very short time the degree of evolution where every possible metabolic step had its eRNA optimized in both efficiency and concentration.

The metabolic network of the prebiotic chemoton shown as an example contained about 10^2 elementary reaction steps; the metabolic network of today's prokaryotes contain about 10^3 reaction steps, all with their own enzymes. This evolution may be explained by the fact that the appearance of eRNAs made possible not only the acceleration of reactions, but also the occurrence of reactions that were not taking place at all (or most only negligibly) in chemical evolution. Thus the appearance of eRNAs reacting with the metabolic network may have started a new phase of stoichiometric evolution; making possible the expansion of the metabolic network by reaction pathways that might not be present in early chemotons. This may have been the phase in prebiotic evolution in which aromatic amino acids, deoxyribose, thymine, etc. appeared.

Let us suppose that the metabolism consisted of, e.g., 5×10^2 different reactions at the time of the evolutional optimization of eRNA chemotons, i.e., they had that many different e-RNAs (according to Eq. [9.13], on average about 500 molecules of individual eRNAs were present in one chemoton). This is a large enough number for ensuring the coexistence and survival of different eRNAs. However, the large variability that has been the driving force of evolution, so far, as it has made possible the appearance of chemotons with shorter and shorter generation times, becomes a disadvantage now that the generation time could not be shortened any more. The probability namely, that the fitness of descendants will be greater than, or even identical to, that of their parent chemotons has decreased. Thus the evolution "strategy" changed, and further on those chemotons had the advantage over the others that changed less, because less of its descendants had a decreased proliferation rate. Thus, following optimization, those chemotons capable of decreasing their degree of mutation through a newly developed, hereditary mechanism were in the most advantageous position.

Variations caused by point mutations could be decreased by the enzymatization of the template process. The hypothesis put forward by Brewin (1972) has already been mentioned, according to which (based solely on structural considerations) RNA molecules, if in an appropriate conformation, may be suitable for polymerase functions. The appearance of eRNAs with polymerase functions decreases the probability of point mutations by orders of magnitude, thus in this phase this represents a driving force in the direction of the spreading and optimization of polymerases.

The probability of replication defects discussed in connection with loop formation (Fig. 9.4) also decreases significantly with the appearance of polymerases. However, mutations originating from the junction of strands remain. They occur with a high probability, as the imidazolides at chain ends are very reactive.

Thus in this phase of prebiotic evolution it was advantageous if the synthesis of RNAs was performed with other, not so reactive nucleotides instead of the very reactive nucleotide imidazolides. Experiments with the prebiotic synthesis of nucleic acids have shown that this is not possible in nonenzymatic reactions. However, in enzymatic reactions the polycondensation of nucleotides can be carried out with both nucleotide diphosphates (polynucleotide phosphorylase) and nucleotide triphosphates (RNA polymerases). Thus it can be assumed that the nucleic acid metabolism changed from one based on nucleotide imidazolide to one based on nucleotide triphosphate (or diphosphate?) at the time when RNA polymerase RNAs were developed.

The appearance of DNA

According to the train of thought outlined above, chemotons optimized in the eRNA phase of prebiotic evolution already had enzymatic metabolism, the structure of enzymes being determined by genes and the information concerning enzymes coded in the nucleotide sequence of genes. The differences – that the enzymes were not proteins but RNAs and that the genes were not DNAs but RNAs – are not of a conceptual, functional nature, but rather only differences in the manner of realization. However, there are some basic, functional differences between the operation of these chemotons and prokaryotes.

First, in prokaryotes (and in fact in today's whole living world), the flow of genetic information is unidirectional; corresponding to the so-called central dogma, i.e., only variations in genes may be transcripted onto enzymes and not vice versa. In chemotons containing eRNAs and gRNAs, the flow of information is bidirectional, i.e., variations in enzymes are necessarily transcripted onto genes just as changes in the genes onto enzymes. This meant a great advantage in the first phase of prebiotic evolution, as it made possible a quick, not accidental development of enzymes and genes. However, according to the previous reasoning, in this phase of evolution the decrease in the variability is advantageous from the point of view of evolution, thus individuals capable of changing the bidirectional flow of genetic information into a unidirectional one by some mechanism are in a more advantageous situation.

Another conceptual, functional difference is that in these chemotons the amounts of enzymes and genes belonging to them are stoichiometrically determined in a ratio of 1:1, i.e., the numbers of enzyme and gene molecules are identical. This relation holds for individual enzymes as well as for the total of all enzymes. In prokaryotes, on the other hand, the number of enzyme molecules and that of genes related to them are stoichiometrically independent. As from one gene sequence any number of readings can be carried out – by means of a suitable reading mechanism – the 1:1 stoichiometric relationship between eRNAs and gRNAs is not the result of some basic principle, but rather of the lack of a suitable reading mechanism. However, it is obviously not economical to produce genes in such excess amounts. Thus the individuals in which a molecular mechanism develops, that makes it possible to produce more enzyme molecules on one gene, acquire an evolutional advantage.

The third conceptual and functional difference is that prokaryotes (and all of today's living beings) can optimize the amount of their enzymes not only in an evolutional process, but also in the individuals even within a fraction of their generation time. eRNA chemotons are not able to do this in spite of the fact that it makes possible a considerable adaptation of the individual to environmental changes, which is also favorable from the evolutional point of view. This feature, however, supposedly developed after the appearance of the translation processes of the protein-synthesizing mechanism, and thus belongs to biological rather than prebiotic evolution.

Neither the 1:1 enzyme-to-gene ratio nor the flow of genetic information can change if the enzyme and gene are templates of each other. In this phase of evolution there is no possibility for transferring enzyme activity to another macromolecule, but the gene function may be taken over by DNA molecules, as the stoichiometric conditions for this (the synthesis of deoxyribose, thymine, and nucleoside triphosphates) may already have been created in the enzymatically expanded metabolic network, and the RNA polymerase supposed by Brewin is just an RNA-dependent DNA polymerase, i.e., a reverse transcriptase catalyzing the synthesis of DNA on an RNA template. If DNA genes were formed in prebiotic evolution as a result of the appearance of reverse transcriptase, the gene sequence was not accidental but rather the eRNA sequences preformed and optimized in earlier evolutional processes could be transcript into the newly formed DNAs.

The appearance of genetic information coded in DNAs had a number of consequences and opened up new possibilities for evolution. First, the syntheses of genes and enzymes became separated, as the double-helix DNA became its own template. But it also became the template of eRNAs through polymerase enzymes, and in a process independent of the replication of DNA. This permits the synthesis of any number of enzymes on a single gene.

The most important consequence is perhaps the change in the mechanism that initiates replication. In today's living systems the separation of the double-helix structure needed for template polymerization occurs via an enzymatic process. In systems of the early stages of prebiotic evolution this was not possible, not only because of the lack of appropriate enzymes, but also because after the formation of eRNAs the chain ends, as a result of their different globular structures, were in different spatial positions with respect to the neighboring chain ends and thus the formation of enzymes capable of dehelicalizing all of the eRNAs was hardly possible. This is why we have assumed, when discussing the prebiotic chemoton, that dehelicalization is regulated by the nucleotide concentration in a way that below a certain concentration RNAs are of a helical structure (in the case of eRNAs they are globular and possess enzyme activity), but above this critical concentration they are dehelicalized and operate as templates.

The filamental DNA structure enables the RNA polymerases to dehelicalize the DNA molecules by themselves in the course of replication. Moreover, on a DNA having a double-helix structure, eRNAs can be synthesized without the dehelicalization of the DNA molecule. When these mechanisms became operative, it was no longer necessary for the nucleotide concentration to increase, above the critical level, as the syntheses of both enzyme and gene could also take place at much lower nucleotide concentrations. This, in turn, means that gene RNAs and enzyme RNAs could no longer be replicated on each other and thus gene RNAs would be eliminated from the metabolism and DNA could become the only carrier of hereditary properties.

The origin of code

The origin of the genetic code has been investigated by numerous authors during the last decades (Balasubramarian, 1982; Barricelli, 1977; Bloom, 1966; Crick, 1966, 1967, 1968; Crothers, 1982; Grafstein, 1983; Hendry and Bransone, 1981; Hendry and Witham, 1979; Ishigami and Nagano, 1975; Jukes, 1965, 1981, 1983; Jungck, 1978; Jurka et al., 1982; Lacey et al., 1975; Macchiato and Tramontano, 1982; Nagyvári and Fendler, 1974; Pelc and Walton, 1966; Reuben and Polk, 1980; Root-Bernstein, 1982; Shepherd, 1981; Shimizu, 1982; Weber and Lacey, 1978; Woese, 1968; Woese et al., 1966; Wong, 1975, 1976) partially based on differing, contradictory assumptions, however, with, only little success. To date there has not even been found a most probable course of the genesis of the genetic code based on these hypotheses. Therefore this very widespread research will not be discussed here; it will suffice to say that if the chemoton theory correctly reflects, at least in its main features, the course of prebiotic evolution, then the origin of the genetic code must be looked for in a way completely different from those studied until now. Just as in the whole course of evolutional processes, earlier events always determined the events that were to take place subsequently; this is also the case for the newly forming code. Just as we did not have to assume accidental events (occurring only with a very low probability) when discussing chemical and prebiotic evolution, we must not consider the possibility of the "frozen accident" concept assumed by Crick, even in the last step of prebiotic evolution, in the genesis of the genetic code.

In this case the driving force seems to have been the enzyme activity already developed and even optimized at the RNA level. It is hard to imagine that the primitive enzyme activity of proteins accidentally produced by a protein-synthesizing system (itself being only in a developing state) could compete evolutionally with optimized eRNAs. A competition can only be imagined if the appearance of proteins could develop further the evolutionally exhausted possibilities of eRNAs and, in particular, not by the random formation of new genes, but by using the already existing ones. However, if on the one hand enzyme functions existed together with the tertiary structure belonging to them (or at least with functional parts of it) and, on the other hand, the genes were provided with evolutionally fixed sequences, proteins could take over the functions of eRNAs only if the genetic code ensured from the very first that the nucleotide sequence of the given gene could be coded over into a protein of an amino-acid sequence provided with the most important spatial characteristics of eRNAs (Gánti,1983)

Based on theoretical considerations and on our knowledge concerning the structure of tRNAs, we may assume that the active sites of eRNAs are mainly their looplike parts. From this it follows that if the function of developing proteins was determined by genes optimized through eRNAs, then their active sites were translated from palindromic sequences. Because active sites are relatively conservative sequences within the polypeptide chains of enzymes, they may possibly also be found in today's enzymes. In other words, it is possible that the parts of today's proteins that are regarded as ancient (i.e., the active and conservative sections) still carry, even today, the information of looplike structures characteristic of primitive eRNAs.

In order to study this, let us set up the "pairing vocabulary" of codon triplets, i.e., let us see which codon triplets of an individual amino acid may pair with the codon triplet of another amino acid. If among the codon triplets of two amino acids there is at least a pair complementary to each other in their sequence, then they are regarded as capable of pairing. This does not mean, however, that the amino acids themselves would be the pairs of each other in some way. The "pairing vocabulary" of amino acids thus obtained is summarized as follows (Gánti, 1983):

Gly/Ala	Ser/term
Gly/thr	Asn/Ile
Gly/Pro	Thr/Cys
Gly/Ser	Thr/Arg
Asp/Val	Ile/Tyr
Asp/Ile	Arg/Pro
Ala/Ser	His/Met
Ala/arg	Glu/Leu
Ala/Cys	Pro/Tyr
Val/Asn	Leu/Lys
Val/His	Leu/Glu
Val/Tyr	Leu/term
Ser/Thr	Glu/Phe
Ser/Arg	Lys/Phe

Table 9.1

When the codon triplets corresponding to the amino acid sequence of every peptide chain are known then, theoretically all the nucleotide sequences that may be the messengers of the given polypeptide can be constructed. However, in practice, in the majority of cases this is can not be accomplished, because the number of messenger variants for a polypeptide chain of a given amino acid sequence is extraordinarily large, especially when considering the fact that an amino acid may even be coded by three or four different triplets. However, because the "pairing vocabulary" directly provides the pairs of amino acids among the triplets of which there are some with complementary character, the amino-acid sequences to which loop-structured messengers may be assigned can readily be found by means of this vocabulary.

We compared the pairing possibilities and the loop-sequences assignable to them in an earlier investigation (Gánti, 1983) by means of the active site sequences of the Dayhoff atlas (Dayhoff, 1972). A sequence was considered to have loop structure if at least two neighboring amino acids were capable of pairing with two other neighboring amino acids in a section of the sequence near them but separated by at least one amino acid. In this case, namely, in the RNA sequence belonging to it, six base pairs are situated adjacently, which already ensures the formation of a stable helical structure.

It is notable that almost every active site sequence may have a loop RNA structure assigned to it and that several of them may have even more than one. Similarly, loop-structured RNAs can be assigned to the conservative sequences of proteins as well. However, these data do not constitute decisive proof. Loop structures can be assigned to the non-conservative sequences of proteins as well, although, according to informative studies, only to a smaller extent. The final decision on this matter can only be made after the study of a great number of sequences and this has not yet been undertaken. However, these studies do show unambiguously that the existing genetic code provides a possibility for turning loop-structured RNA sequences into protein sequences of various active site functions.

Summary

The development of the genetic code may be considered the beginning of biological evolution, as a proliferating and evolving system carrying its hereditary properties in the base sequences of its DNA, which determine the properties of protein enzymes when protein enzymes regulate metabolism, belongs indisputably to the frame of biological evolution. Thus by means of the chemoton theory it seems to be possible to provide a uniform, comprehensive answer to the most unknown phase of the evolution of matter: to the spontaneous genesis and subsequent prebiotic evolution of living systems.

It must be repeatedly emphasized that we do not state that events occurred exactly in the way presented. A large part of the processes described are based not on proven facts but rather on a logical chain of hypothetical events. However, there are certain facts that may be stated:

1. The genesis of life was not accidental, but a necessary event taking place not only once by chance, but at any time and in every place where the preconditions for its occurrence were given.

2. The problem of the origin of life cannot be solved without knowledge of the essence of life. No mechanical engineer can design a machine; in general, he has to know which kind of machine he wants to design and what the essence of this machine is.

3. The chemoton theory provides one possible solution to the question, what is the essence of life? By means of it living systems may be projected and possible ways of biogenesis and prebiotic evolutional processes may be elaborated. One of these possible ways of biogenesis together with the prebiotic evolution connected to it has been outlined in the last two chapters. The main features of ways provided here are in accordance with our current experience gathered.

4. The chemoton theory and the *cycle* stoichiometry and stoichiokinetics developed within its framework enable us to quantitatively describe the chemical events of these processes.

5. The chemical network of processes given has some very remarkable properties:

   a) It is the first to provide a possible description of the whole series of events, i.e., starting from the simplest cosmic compounds it builds up step by step the evolutional chain leading to prokaryotes.

   b) The steps form a logical chain; every step is a logical consequence of the preceding one.

   c) The steps are canalized. Almost every step of the organization offers various, sometimes very many solutions, but this variability is decreased by the events of the preceding organizational level and they make the occurrence of one specific event a logical necessity. Because of this canalizing effect, improbable and impossible events fall out of the series of events and thus the multi-step evolutional chain from cosmic compounds to prokaryotes becomes a chain of necessary events.

d) Though this multi-step evolutional process is far from being proven, experience and/or experimental facts supporting it may be assigned to each of its steps.

e) It contains numerous predictions some of which (e.g., the existence of eRNAs) have in the meantime been proved.

f) Finally, it enables the researcher to pass over many of the different experimental ways suggested so far.

X. Exobiological Outlook

Introductory Remarks

In what follows, exobiology refers to the science dealing with extraterrestrial life, regardless of whether the given biological object is of terrestrial origin or not. Though its roots are to be found in ancient times (e.g., in the letters of Empedocles) and have surfaced again and again since then (e.g., in the works of Giordano Bruno), this field of science is even now still in the process of birth, for no living systems of non-terrestrial origin have as yet been found. However, experience originating from terrestrial experiments and, as a result of space research, data concerning extraterrestrial objects are accumulating. This may support exobiological speculations and transfer this subject more and more from the world of science fiction into the world of the exact natural sciences.

At present, not even the nomenclature of exobiology is uniform; terms such as astrobiology, bioastronomy, cosmobiology (coupled awkwardly with astrology), biocosmology, space biology, planetary biology, intellexobiology, etc. are used (Freitas, 1983). This subject is sometimes even expanded to include questions of communication with hypothetical extraterrestrial civilizations, which surely fall outside the frame of exobiology (though they are not independent of it, as the precondition of a civilization is the existence of an appropriately developed living world).

The discussion of extraterrestrial civilizations is beyond the scope of this book. This chapter has been included not because of a fascination with the subject, but because the chemoton theory may provide general theoretical concepts not available before. Exobiological speculations tend to be rather earthbound, as experience to date originates from terrestrial living beings. As an example, let us see what planetary and "protobiological" conditions are set in the best, most recent survey predicting the development of an extraterrestrial living world (Oró, Rewers, and Odom, 1982).

Planetary conditions:

1. *Mass of planet.* Sufficient to retain volatiles of C, N, O, S, P, but not so excessive as to retain large amounts of H_3.
2. *Orbit.* Optimum distance of planet to central star continuously provides the necessary energy to maintain a temperature from 0 °C to 100 °C on the surface of the planet.
3. *Atmosphere.* Sufficient to hold the volatiles of C, N, O, S, P, and to protect emerging biochemical molecules and life against solar ionizing radiation and UV.
4. *Discrete liquid sphere.* To provide a medium for trapping molecules and ions (hydrosphere). Synthesis of biochemical monomers.
5. *Surface.* Solid surfaces to provide the boundaries for localized environments. Evaporation and condensation of monomers.

Protobiological criteria:

1. *Replication of information-carrying molecules.* Ability to store information in a simple linear code that is capable of undergoing template replication and may be decoded.
2. *Stereospecific catalytic molecules.* To speed up the synthesis of biochemical compounds inside the microenvironment.
3. *Information transfer molecules (hybrids).* To transfer information from code to catalytic molecule.
4. *Polymerizing molecular assembly.* To synthesize catalytic molecules by template-directed polymerization.
5. *Molecular assembly between phases.* To provide an appropriate microenvironment and increased internal reaction rates.

It is obvious that these criteria reflect the present terrestrial living world and in fact are even too restrictive for that. For example, the upper temperature limit of 100 °C is derived from the atmospheric pressure on the surface of Earth, but even on Earth there are living beings capable of proliferation above 100 °C. Neither can the filtration of the ionizing radiation of the Sun be a criterion, for it is known that, in the primary and secondary cooling cycles of nuclear reactors, microorganisms are living and proliferating in spite of the fact that they are exposed to extremely intense ionizing radiations (Hortobágyi and Vigassy, 1966, 1967). Increased UV radiation would not make life impossible in water or soil, not even under terrestrial conditions. The above listed protobiological criteria entirely reflect the separation of the coding and catalytic systems (i.e., the present functional separation of DNA–RNA–protein), which is not a criterion of life even under terrestrial circumstances, as has already been discussed, even if the definition of life were confined to the presence of specific catalysts.

Chemoton theory may provide help to exobiological research in two aspects. First, the pathway outlined in Chapters VII and VIII for the origin and early evolution of life may be used for the exact study of the non-terrestrial origin of systems with a terrestrial type of metabolism. Second, and this is perhaps the more important aspect, the abstract chemoton model may provide a theoretical basis on which to decide whether the existence of living systems with a non-terrestrial metabolism or perhaps that non-carbon-based systems may be a real possibility, because the model binds the essence of living systems to their functional properties rather than to the chemical nature of their materials. Thus its equations remain valid for the reactions of any chemical compounds so long as they are capable of satisfying the functional criteria determined by the equations.

Possibility of the Existence of Non-terrestrial Types of Life

Non-carbon-based life

This possibility is dealt with mainly in science fiction. In scientific literature, at most the possibility of silicon-based life is considered, and rejected. However, rejection is based not on its unsuitability for building living systems, but rather on the comparison of carbon and silicon atoms, stating that carbon atoms have some properties utilized in living systems that are not present in silicon atoms. For example, carbon atoms are capable of forming double and triple bonds with other carbon atoms as well as with atoms of a different type, whereas silicon atoms form mostly single bonds. Carbon atoms can form long chains and rings, whereas from silicon atoms only short chains can be formed, though they have the capability of chain formation (a maximum of eight silicon atoms can be bound together).

According to several authors, the main difference between carbon and silicon atoms with respect to life is that compounds containing Si–Si bonds are very labile, e.g., the silicon analogue of ethane, disilane, which has the composition $H_3Si\text{-}SiH_3$, reacts even at ambient temperatures with water by forming silicic acid and hydrogen. Thus no compounds containing Si–Si bonds occur on Earth. There is also a difference between their redox properties, e.g., CO_2 forms very diverse redox products in photosynthesis, whereas SiO_2 can only be reduced at temperatures of several thousand degrees centigrade, the product always being elemental silicon.

However, these experiments only prove that in terrestrial life, carbon atoms cannot be substituted by silicon atoms. There is no theoretical basis for stating that life cannot exist without compounds containing double and triple bonds, or that in living systems only macromolecules consisting of chains formed from identical atoms may exist. In fact, if silicon atoms are bound with other atoms, e.g., by oxygen atoms, diverse macromolecules can also form from silicon atoms. Although redox processes are the main driving force of terrestrial life, at present there is no theoretical restriction that the energy of nonredox processes cannot play the role of a driving force in living systems. This cannot be said even though we were taught in school that the most important energy transfer process is the ADP–ATP transformation, which, as is well known, is not a redox reaction. Though it is true that the metabolism of terrestrial living systems is based on redox processes, this is only a consequence of the fact that during chemical evolution redox reactions prevailed on Earth, owing to the composition of the primitive atmosphere.

Neither can the argument that compounds with a Si–Si bond are too labile under terrestrial conditions be accepted. If on some planet the conditions are similar to those on Earth, a living world with metabolisms identical to those on Earth must exist, for these terrestrial conditions are created and maintained just by the metabolisms of this living world! Life based on silicon – if it exists at all – would be formed and maintained under conditions totally different from those on Earth!

Another argument often mentioned is that in the distributions of the frequency of elements in the Universe – noble gases excluded – the biogenic elements head the list. This is also mistaken because it is not the average frequency of elements but rather their local concentrations that are decisive; there are numerous data already available that clearly illustrate the following: in the condensation of matter in the Universe a series of processes are operative that lead to an extraordinary local enrichment of several elements or element groups. There are numerous surprising and excellent examples for this even within our solar system, some of which will be dealt with later.

The truth is that the "scientific" arguments in exobiology are totally unfounded from the biological point of view, for the current biology has experience only with terrestrial life and not with life in general, and no one has even tried to extrapolate this experience to a general life theory. The chemoton theory provides a possibility for an exact advancement in this field also by considering living systems as fluid (chemical) automata, by studying their general organizational mode, and by describing this mode of organization in an abstract way. On studying the abstract metabolic map of the chemoton model, the cyclic equations providing its stoichiometric relationships, or the kinetic equation systems describing its time course, it should be noted that there is no single component in the model with a predetermined chemical composition. The relations thus obtained are true regardless of whether compounds based on carbon, silicon, sulfur, nitrogen, phosphorus, or any other element are substituted into the equations instead of the letters, as long as they are capable of performing the reactions determined by the metabolic map or the stoichiometric equations of chemotons. Thus the following question can be unambiguously studied in an experimental fashion: is it possible to construct, from non-carbon based compounds (but perhaps also containing carbon), systems of reactions satisfying the organizational requirements of chemotons, and can they be formed under conditions different from those on Earth?

What are the criteria of the chemoton theory?

1. Living systems are fluid (chemical) automata, thus for their operation they need a liquid phase.

   In the thinking of biologists, the liquid phase is uniformly water, whereas in science fiction celestial bodies with seas of methane, ammonia, sulfuric acid, hydrogen fluoride, etc., are also imagined. Until recently, these belonged to the realm of fantasy. However, with the reports that our nearest Sun-side neighbor, Venus, has cloud-forming sulfuric acid droplets as its only liquid phase (Prinn, 1985), that the moon Io, of our second neighbor on the opposite side, has sulfur seas on its surface – an idea inconceivable even in science fiction (Johnson and Soderblom, 1983) – and that on the giant moon, Titan, of our next neighbor Saturn there are methane–ethane seas and methane rains fall on its surface, it is not impossible to imagine celestial bodies with seas of ammonia or hydrogen fluoride.

2. One of the subsystems of living systems should be a self-producing (autocatalytic) cycle (network).

   At present, we know very little about chemical processes taking place in a hydrogen fluoride sea (in which silicates can be dissolved), in a sulfur sea, or in an ammonia sea under the conditions prevailing there. However, it is obvious now that the effect of radiation energy from outside the celestial body may result not only in a chemical network consisting of terrestrial-type geobiochemical organic compounds, but also in some other organic networks (as in the case of Titan, which will be dealt with later), and even in inorganic networks; for example, in the atmosphere of Venus. These networks also include catalytic cycles: e.g., chlorine oxides (ClO_x) play a catalytic role in the atmospheres of both Earth and Venus (de More and Young, 1982).

Thus the second criterion cannot be either accepted or rejected on the basis of our present knowledge. However, this problem can be studied, by laboratory experiments, in an exact and unambiguous manner; after having made the appropriate experiments, it can also be solved. A positive outcome to such experiments seems to be the more probable, as autocatalytic networks are also known in inorganic chemistry (e.g., among the reactions of H_2O_2, iodates, bromates, and chlorates).

3. Living systems should include subsystems suitable for information storage and template polymerization.

Theoretically, every macromolecule that is built in a non-periodical sequence may be suitable for storing information. Some theoretical possibilities for non-carbon-based systems are shown in Fig. 10.1. In case a) the chain length and arrangement of S_1, S_m, S_n, etc. may carry information, whereas in case b), in addition to chain lengths, groups R_1, R_2, R_3 (which may also contain carbon atoms) and their sequences are information carriers. The situation is similar for cases c) and d). Supposedly, case e) is the least acceptable for chemists. However, they should be reminded that a compound, glyoxal-bis(guanylhydrazone), with the composition

```
HN = C – NH – N = CH – CH = N – NH – C = NH
     |                                 |
    NH2                               NH2
```

exists in which the N/C ratio is equal to 2, and the existence of diimine corresponding to ethylene with the formula

```
       H
      /
   N = N
  /
 H
```

can also be detected at low temperatures, suggesting that at those temperatures the formation of nitrogen-containing polymers is imaginable ("Exotic Biochemistry in Exobiology," 1970).

These are, of course, only hypotheses for non-carbon-based polymers capable of information storage, and experimental data supporting them are still missing. Nevertheless, these problems can also be clarified by terrestrial laboratory experiments. However, a molecule suitable for storing information is not necessarily at the same time a molecule capable of template polymerization. Then again, the experiments of Weiss concerning the formation of several clay minerals by replication using the technique of template polymerization also suggest that template polymerization is not a privilege of nucleic acids or fiber polymers (Weiss, 1981). This question can also be decided upon by laboratory experiments under terrestrial conditions.

4. Living systems should be surrounded by two-dimensional liquid membranes.

The major question concerning this criterion is whether silicon, sulfur, nitrogen, or other non-carbon-based compounds can form two-dimensional fluid membranes. Experimental data are again missing here, but the experiments of Herrera (Herrera, 1928, 1934, 1942) together with

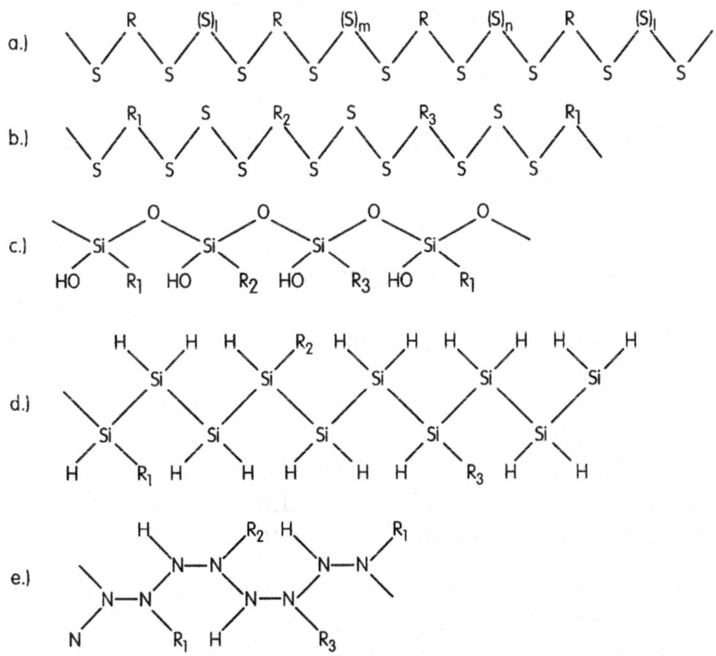

Fig.10.1. Some principal possibilities of chemical information storage on a non-carbon basis.

the "osmotic fungi and algae" of Leduc (Leduc, 1907, 1910), as well as other osmotically growing structures suggest that the formation of such membranes cannot be excluded. This is yet again a question answerable by suitable laboratory experiments.

5. Subsystems listed in points 2–4 should be capable of entering the stoichiometric relationship required by the chemoton model.

   When possession of experimental data answering the questions rose in points 2–4 is attained, this problem can be solved, even at the theoretical level. That is, if we know the chemistry of non-carbon-based reaction networks, the concrete nature of their autocatalytic cycles, template polymerization processes, and membrane formation, then a concrete, non-carbon-based system with chemoton coupling can be constructed just as has been done for carbon-based prebiotic chemotons from the products of chemical evolution.

   Thus the chemoton theory cannot provide an answer as yet for the question, whether a non-carbon-based living system could exist, because of lack of appropriate experimental data. It moves this problem from the world of fantasy into the field of exact, experimentally yet unsolvable questions, which may once be answered experimentally.

Carbon-based life with a non-terrestrial type of metabolism

In every place where organic matter is associated with a liquid phase of water, the existence of a terrestrial type of life can be expected. However, in places where organic matter together with an appropriate source of energy is present, but where the liquid phase is not water, no terrestrial type of metabolism (carbohydrates, organic acids, etc.) can be imagined. Until recently, such circumstances belonged only to the realm of science fiction. However, from the studies of the spacecraft Voyager it is clear that on Titan, the largest satellite of Saturn, the sea consists of a mixture of liquid methane and ethane, and methane, ethane, or methane–ethane mixtures precipitate depending on the height of the clouds (Sagan and Dermott, 1982; Eshelman, Lindal, and Tyler, 1983; Flasar, 1983; Lunine, Stevenson, and Young, 1983; Owen, 1984a, b, Raulin, 1984).

The main components in the atmosphere of Titan are nitrogen, argon, and methane. Because argon is a noble gas, incoming energy is absorbed by the CH_4–N_2 system. The result of energy absorption in such systems has been studied by numerous authors; it is dealt with in more detail, including a literature survey, in the work of Raulin, Monrey, and Toupance (1982). From their work it is obvious that the result of energy absorption in such systems is the production of unsaturated hydrocarbons and cyano derivatives. These are very reactive compounds capable of polymerization even under cold conditions, transforming this fact, the carbon compounds of Titan should be accumulated on its surface as polymers.

However, a study of the atmosphere of Titan has shown that besides nitrogen, it also contains a significant amount of methane; ethane and propane are present in smaller amounts, whereas unsaturated carbon and nitrogen compounds occur only in traces. This is illustrated in Table 10.1 (Sagan, 1983). Unsaturated carbon and nitrogen compounds should also have been polymerized with time under the conditions on Titan, and their extraordinarily low concentrations suggest that they react in parallel with their formation. If there were no processes transforming the products back into methane by decomposition, methane should have vanished from the atmosphere of Titan.

Component	**Quantity**
Nitrogen (N1)	82-99%
Argon (Ar)	0-12%
Methane (CH_4)	1-6 (it changes with height, the total amount is uncertain)
Hydrogen (H_2)	2000 ppm
Acetylene (C_2H_2)	2 ppm
Ethylene (C_3H_4)	0.4 ppm
Ethane (C_3H_6)	20 ppm
Diacetylene (C_4H_2)	0.1-0.01 ppm
Methylacetylene (C_2H_4)	0.03 ppm
Propane (C_2H_6)	20 ppm
Cyanogene (C_2N_2)	0.1-0.01 ppm
Hydrogen cyanide (HCN)	0.2 ppm
Cyanoacetylene (HC_3N)	0.1-0.01 ppm
Carbon monoxide (CO)	50-150 ppm
Carbon dioxide (CO_2)	0.0015 ppm

1. Table 10.1. Composition of the atmosphere of Titan (Sagan, 1983)

On Earth, this decomposition is performed by living beings, presently into the oxidized basic compound of carbon, CO_2. It can be assumed that, as was mentioned earlier when discussing inverse assimilation, in the primordial atmosphere this back-transformation occurred via methanogenic bacteria. For this reaction H_2 is needed. In the atmosphere of Titan, H_2 is present only in very low concentrations. However, this back-transformation cannot be performed by terrestrial types of methanogenic bacteria, as their metabolism operates only in an aqueous medium. There are two possibilities left: back-transformation is carried out either by abiogenic chemical cycles, or by living beings having a non-terrestrial type of carbon-based metabolism. The possibility for the genesis of this type of living being may also be studied by means of the chemoton theory, with exact methods and in laboratories under terrestrial conditions, just as was mentioned concerning non-carbon-based systems.

Life of non-terrestrial origin with a terrestrial metabolism

Experimental data collected to date seem to prove that on celestial bodies with water as their liquid phase, the living world would be as one with the terrestrial type of metabolism. However, in this case the "terrestrial type of metabolism" does not necessarily mean a flora producing O_2 and consuming CO_2 by photosynthesis, a fauna based on respiration as the main source of energy production, and not even a specific catalytic regulating system based on proteins. Photosynthesis as coupled with water decomposition and respiration did not exist on Earth 3 billion years ago; thus metabolic pathways connected to these phenomena could not exist either, though these are the decisive processes of today's living world on Earth. If the assumption of the chemoton theory concerning prebiotic evolution is correct, the living world must have had a period in which proteins did not play the decisive role. Thus the phenomena of the metabolism of the living world now considered the most characteristic (photosynthesis, respiration, proteins, DNA) are the products of evolution built onto the autocatalytic carbohydrate–organic acid reaction network, the ribonucleic acid metabolism, and the glycerol ester- and/or ether-based membranes during the course of evolution. In fact, the latter are metabolic processes really characteristic of terrestrial living beings.

The statement, that life on "wet" celestial bodies is supposedly always of the terrestrial type of metabolism means, that their most general metabolic processes are similar to those on Earth. This may be stated because their formation is unequivocally determined by thermodynamic parameters and, according to chemical evolution experiments, in the presence of water the reactions tend to proceed toward the formation of the components of terrestrial metabolism, even if in the concentrations of the individual components significant differences may be observed depending on the conditions.

However, the evolutional pathway is determined not only by thermodynamic parameters, but also by the prehistory of the "infinite" series of ancestors. In turn, for every living world this prehistory consists of an infinitely large number of accidental events, and is influenced also by the canalizing effect of selection caused by external circumstances. Therefore the evolutional pathway cannot be predicted; at most only those possibilities may be revealed that are open to a living world in the course of its evolution. However, the magnitude of such possibilities is so high that at best several general tendencies may be suggested, such as that the evolution of a living world proceeds toward complexity, whereas that of the individuals moves toward specialization, or that evolution strives to fill every possible ecological niche, etc. However, these general trends do not give an answer to the question as to in what respect a life with a terrestrial type of metabolism but not of terrestrial origin would differ from the living world developed on Earth.

Exobiological Studies in the Solar System

Dry celestial bodies

In this context, dry celestial bodies are those not having water on their surface, but on which other liquid compounds may be present. No liquid phase at all exists on Mercury, on some of the moons, on small planets, and on meteorites. On these, there is obviously no life possible. Nevertheless, exobiological studies have been carried out in connection with them insofar as, for example, rocks from the Moon and the material of numerous meteorites have been examined for organic matter.

In both the rocks from the Moon and the carbon-containing meteorites, the so-called chondrites, numerous organic compounds were found, though in very small amounts. These were formed in the non-aqueous phase of chemical evolution and from them organic compounds of biological importance could be formed in an aqueous medium. Thus it may be concluded that at least in our solar system (but, according to earlier mentioned radio-astronomical measurements, also outside it), chemical evolution toward life has started everywhere. However, lacking an aqueous phase, evolution in most cases has been halted at the formation of unsaturated organic compounds.

Venus was earlier thought of as the aqueous twin of Earth: it is not much smaller than Earth and its surface is covered by clouds. However, the spacecrafts Venera and Pioneer provided surprising data about Venus: its surface temperature is higher than 400 °C, its atmospheric pressure is 90 atm, the atmosphere consists of almost pure CO_2, and the clouds are not of water but of sulfuric acid droplets.

Under such conditions no form of life can exist, though Morowitz and Sagan (1967) suggested that, in the heights of the clouds, balloon-like floating living beings filled with hydrogen may perhaps exist; they would float at such a height that temperature and pressure would be commensurable with those on Earth. However, at the time of this suggestion it was not known that the clouds consisted of sulfuric acid.

The temperature of Venus is significantly higher than would follow from its distance from the Sun. This is attributed to the greenhouse effect of its atmosphere. The greenhouse effect is a consequence of the very high CO_2 content; in the atmosphere of Venus there is about as much carbon as in the Earth's atmosphere, carbonate minerals, and sediments all put together. There is no water at all on the surface of Venus, and even the amount existing in its atmosphere is very small. At the same time, the deuterium content of the atmosphere is about a hundred times higher than would be expected from the hydrogen content. This high D/H ratio is explained by the hypothesis that earlier there were extensive oceans on the surface of Venus, but water was evaporated by the temperature increase, and water molecules were decomposed into hydrogen and oxygen. The oxygen thus formed then oxidized the organic compounds to CO_2, whereas the hydrogen escaped into space, although its heavier isotope, deuterium, was retained in the atmosphere (Donahue et al., 1982; McElroy, Prather, and Rodriguez, 1982).

If this hypothesis is true, then after the development of the solar system, very similar conditions may have existed on Venus and Earth, thus a living world would have also developed on Venus, as would follow from the chemical and prebiotic evolution on Earth.

Io, the innermost among the so-called Galilean moons of Jupiter, has a very peculiar gravitation; its size (**r** = 1815 km) is almost identical with that of our

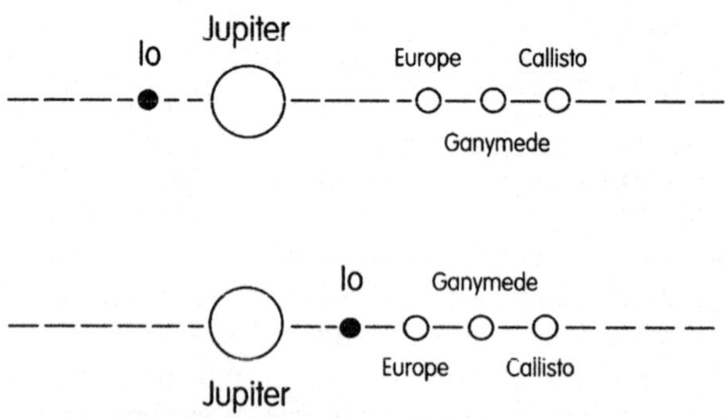

Fig. 10.2. Relative variations in the position of the Galilean moons of Jupiter.

Moon ($\mathbf{r}$ = 1738 km), but its moon neighbors on outer orbits are even larger ($\mathbf{r}_{Europa}$ = 1569 km, $\mathbf{r}_{Ganymede}$ = 2631 km, $\mathbf{r}_{Callisto}$ = 2400 km), thus moons and Jupiter attract it, in one extreme revolution position, from the same direction; in the other position, from opposite directions (Fig. 10.2). Therefore the gravitational field acting on Io changes continuously, and so strongly that the heat produced by friction resulting from the tidal movement inside the moon warms it up. However, at this temperature the mass of Io is not enough to retain the volatiles, thus Io has lost its water, nitrogen, and carbon compounds. On the other hand, internal heat production melted sulfur and sulfuric compounds and kept them in the liquid phase, and this fluid then broke to the surface. Thus above the solid surface an ocean consisting of molten sulfur and sulfur dioxide covered with a layer of solid sulfur and sulfur dioxide developed.

The continuous internal heat production leads to periodic volcanic eruptions in which sulfur and sulfur dioxide are shot out with a speed of up to 1000 m/s and fall back again in an umbrella-like fallout form. The diameter of such fallouts can reach 1400 km in larger eruptions, whereas in smaller ones it is about 300 km. The temperature of the eruptive material is in larger eruptions about 650 K, in smaller ones about 400 K (Beatty, 1979; Johnson and Soderblom, 1983).

Thus our nearest and, up to the present, only opportunity for studying the possibility of an eventual non-carbon-based life form is provided by Io. Though a significant flux of energy arrives at its surface, and its own energy production is even larger, in order to utilize this energy for any life function it should be converted into chemical energy thereby providing a chemical potential difference between the components of the liquid phase, which could serve as the driving force of an appropriate metabolism. Our present knowledge does not provide any basis for such a transformation of solar energy or internal thermal energy. Thus we have to say that, according to our present knowledge, no form of life can be expected to exist in the sulfur seas of Io. On the other hand, this possibility is real for Titan, the giant (Mercury sized, $\mathbf{r}$=2575 km) satellite of Saturn. The UV radiation of the Sun excites methane; therefore in Titan's atmosphere unsaturated carbon and nitrogen compounds of high energy content are continuously forming, which may be suitable energy sources for a living world. The surface conditions on Titan have already been described; here it will only be noted that if life is possible at all with methane as liquid phase (this would be a life of non-terrestrial metabolism), it would probably be found on Titan. Nevertheless studies concerning this hypothesis will probably keep us waiting for several more decades.

Mars

Mars is a dryer planet than was supposed previously, but it is not totally dry. It is the only celestial body that was searched for life of a non-terrestrial origin by spacecrafts (Soffen, 1976). This is understandable, for whenever extraterrestrial life has been discussed, it was always Mars that was considered because its poles are covered with ice caps that are subject to seasonal changes; therefore on its surface the presence of liquid water can be expected. Another reason was the well-observable variations of colors on its surface, which were supposed to correspond to seasonal changes of the vegetation on Mars. Mars was considered the "older brother planet" of Earth and was supposed to have surface conditions similar to those on Earth.

Among spacecrafts searching for traces of life on Mars, Viking-1 (VL-1) landed on July 20, 1976, at latitude 23, Viking-2 (VL-2) 6 weeks later on the opposite side of Mars, at latitude 48. These landing sites would correspond under terrestrial conditions to the geographical positions of the Hawaii Islands (VL-1) and Budapest (VL-2), respectively. Biological measurements were carried out by VL-1 for 7 months, by VL-2 for $8\frac{1}{2}$ months, but other measurements were made for much longer times, e.g., VL-1 transmitted pictures and metrological data for 2245 Martian days (= 3.3 Martian years) (Arvidson et al., 1983). One Martian day corresponds to 24 hours and 39 minutes.

The lack of a general life theory showed unfavorably mostly in the phase of designing the experiments: a very great uncertainty prevailed as to the detection of the existence of an extraterrestrial living world. The designers decided finally to try to detect living systems with a terrestrial type of metabolism, which is justified, but in some details such earthbound devices were used that they interfered with the realization of the measurements (e.g., the presence of vitamin C in the medium in one of the experiments).

In addition to terrestrial astronomical observations, the measurements of spacecrafts Mariner and Mars were also available for use in the design of the experiments. Based on these data, Mars was supposed to be a planet with a rare atmosphere, in which the surface temperature rises above 273 K over significant areas during the Martian summer, and in which, if only in small amounts and temporarily, liquid water is present. About the atmosphere it was assumed that it is neutral in nature, consisting mainly of CO_2 and thus being neither reducing nor oxidizing. Because of the above conditions life could only be expected in the soil, therefore every biological experiment was restricted to the study of the Martian soil.

When designing the experiments, researchers agreed that the studies had to be based on the most general properties of living beings. Thus experiments were aimed at detecting organic materials in the soil and measuring the components (being mainly in their gaseous phase) eventually consumed or produced in metabolic processes.

In selecting the landing site, the main concern was to ensure a secure landing, i.e., the presence of a relatively smooth surface, together with a relatively high temperature (possibly near the equator) and the assumed presence of liquid water (the borderline of an ice cap). This last requirement could not be fulfilled, as no place could be found in the neighborhood of the ice caps that would have been suitable for a secure landing when the close-up pictures of the surface of Mars were examined. Therefore the program of VL-2 was modified before landing, and thus the two modules landed at two different points on the planet (Chryse Plain and Utopia Plain), but essentially at similar types of places, which was a great loss from the point of view of scientific results (Shorthill et al., 1976a,b).

On the surface of Mars, fluvial-like channels of significant size are to be found. These are of two types: one of them may originate from the local melting of permafrost as a result of volcanic activities, whereas the other one is presumably of fluvial origin stemming probably from some atmospheric precipitation. However, meteor craters formed later on the top of riverbeds seem to prove that these traces of flowing water originate from the early period of Mars's history, and thorough analysis of these riverbeds shows that they are formed by eventual one-time flows and not by a continuous flowing of water. This flow could, however, be carrying vast amounts of water. The age of the different riverbeds is estimated at 0.5– 3.5 billion years (Masursky et al., 1977).

There is no flowing water on Mars now. The atmosphere contains only very small amounts of water vapor the local distribution of which changes, whereas the global total vapor content remains constant (Farmer, Danes, and Holland, 1977). The water content of polar areas appears as water ice with varying thickness, which is on average several millimeters thick, but in some places reaches a thickness of up to several hundred meters. In spite of all these facts, Mars is an extremely dry planet compared to Earth because of the very low atmospheric pressure and temperature conditions.

Temperature data measured by the orbiter units vary between 130 and 290 K (Kieffer et al., 1977), whereas those measured by the lander units showed characteristic diurnal fluctuations (Fig. 10.3), but the values were never above the freezing point. On the other hand, the temperature of the collector head of

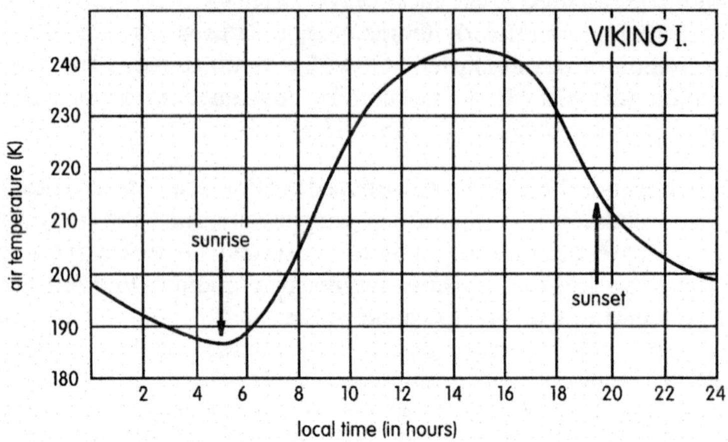

Fig. 10.3. Daily temperature changes on the surface of Mars measured by Viking I.

VL-2 reached 273 K in the soil; thus it cannot be excluded that near the surface, liquid water may eventually appear. In the equatorial regions, the formation of a water-ice sheet is possible at Martian dawns (Hess et al., 1977; Moore et al., 1977).

Both landing sites are in desert regions covered by fine-grained but sticky sand with formations characteristic of wind-blown dunes. During the 3-year observation period, these formations were rearranged by the wind. The strength of the wind can reach 30 m/s (Arvidson et al., 1983). The support of VL-1 penetrated 16.5 cm into the soil. The sand consisted mainly of silicate particles mixed with oxide and carbonate minerals, more or less cemented together with sulfate-rich minerals. The surface was rich in iron, supposedly in the form of hematite (Fe_2O_3) (Baird et al., 1976; Taulmin et al., 1977).

The composition of the Martian atmosphere (Table 10.2) differed significantly from that expected (Owen et al., 1977). The presence of ozone is very important from the biological point of view, because it is a very powerful oxidant and thus keeps the surface particles of the soil in a strongly oxidized state. This is a fact that interfered with the program of biological experiments planned.

Meteorological conditions showed great variations according to the sites. At the landing site of VL-1, the maximum daily temperature measured was -20 °C and the minimum -90 °C, the average being -60 °C. At the landing site of VL-2, the daily winter maximum was -70 °C, and according to calculations, on winter nights the temperature may even sink to -120 °C (Hess et al., 1976a,b).

Though the planet is in general cooler than supposed earlier, in the polar regions the temperature rises above the evaporation temperature of CO_2; thus the hypothesis that the poles contain large amounts of solid CO_2 that serve as a buffer vessel for atmospheric CO_2 cannot be accepted. However, in winter, on top of water ice, CO_2 snow may be precipitated.

Organic materials were sought in the Martian soil by the GC-MS method. This technique was used earlier for determining the organic material content of meteorites, moon rocks, and moon powder (Biemann, 1974). The method is extremely sensitive: the majority of components can be measured in parts per billion amounts. Martian soil (0.1–0.2 g) was heated to 500 °C; the gases and decomposition products evolved were fractionated by gas chromatography and identified by mass spectrometry. Both landing modules could carry out two measurements. One surface soil sample and one sample taken from under the surface were studied by each one. The instruments operated excellently in both modules, as demonstrated by the fact that they also detected the very small amounts of their own technological impurities. Nevertheless, besides these impurities they detected only water and CO_2 (Biemann et al., 1977).

Thus the experiments showed unambiguously that in the soil of Mars, at least at the landing sites of VL-1 and VL-2, there is no organic material at all, or at least orders of magnitude less than in either the chrondrites or Moon rocks. It should be noted, however, that in spite of the high sensitivity of the instruments used, the presence of living systems in the samples cannot be excluded. The mass of a bacterium is about 10^{-12} g, 25% of which is organic material. Supposing that 100

CO_2	95.32 %
N_2	2.7 %
Ar	1.6 %
O_3	0.13 %
CO	0.07 %
H_2O	0.03 %
Ne	2.5 ppm
Kr	0.3 ppm
Xe	0.08 ppm
O_2	0.03 ppm

Table 10.2. The composition of the lower part of the Martian atmosphere.

different compounds are fractionated from the bacterium in the course of a chromatographic analysis, each component has a mass of about 10^{-14} g. Therefore this method cannot detect 1-10.000 bacteria on the basis of quantitative examination of their components per gram of soil, if the soil does not contain any other organic material than that of the bacteria (Gánti, 1978d). Based on the measurements it has to be assumed that in the soil samples originating from the desert region of Mars, all of the organic materials were oxidized to CO_2. For this process, there are several mechanisms suggested.

The essence of the so-called pyrolytic release (PR) experiments has been to study whether gaseous carbon compounds (CO_2 and CO) are capable of incorporating into the soil in the form of organic materials as a result of irradiation or without it. The remote-controlled experiments could be carried out in several ways: with or without a previous (heat) sterilization of the soil sample, with or without irradiation, and with or without wetting (Horowitz, Hobby, and Hubbard, 1976; Horowitz and Hobby, 1977).

In the actual experiments 0.25 cm^3 of Martian soil was weighed into cells with a volume of 4 cm^3. The cells had lids transmitting radiation only above a wavelength of 320 nm, and, if needed, could also be irradiated by a high-pressure 6-W xenon lamp. If necessary, 80 μg of water vapor could also be injected. The gas composition in the cell was 20 μl of $^{14}CO_2$ + ^{14}CO in a volume ratio of 98:2, in addition to the Martian gases present.

The samples were usually incubated for 120 h with or without irradiation and in a wet or dry state. After incubation, the radioactive gases were aired out and the soil sample heated to 625 °C. The volatile products of pyrolysis together with the un-reacted $^{14}CO_2$ and ^{14}CO were transferred to a column that separated them into two fractions: peak 1 containing CO, CO_2, and CH_4 if any, and peak 2

Sample	Experimental conditions	Count/min
Viking-1		
1	irradiated, dry	843±29
2	irradiated, dry, 175°C	105±29.5
3	irradiated, dry	214±28
4	irradiated, dry	289±31
5	irradiated, H_2O, 90°C	275±29
6	irradiated, H_2O	225±31
Viking-2		
1	dark, dry	178±31
2	irradiated, H_2O	-7±28
3	dark, dry	36±35

Table 10.3. The results of PR experiments.

containing organic fragments larger than methane. After that, the column was heated to 640 °C, and because CuO oxidizes organic materials into CO_2 the activity of CO_2 was measured. The results are summarized in Table 10.3. These results cannot be considered positive from a biological point of view, as samples from VL-1 sterilized by heat had also shown "positive" results, whereas the results of VL-2 are contradictory and thus not appraisable.

Gas exchange (GE) experiments have investigated whether gases are evolved or consumed upon the introduction of nutrient media containing organic materials, which can be assimilated by terrestrial microorganisms, into the Martian soil samples.

Experiments were carried out as follows: 1 cm^3 of sieved Martian soil was introduced into cells of 8.7 cm^3 volume. The cell was then rinsed with a neutral gas mixture of the composition: 91.65% He, 5.51% Kr, and 2.84% CO_2. The cell was filled to a pressure of about 200 mbar, and then the experiments could be realized in three different ways:

a) In a dry state;

b) By placing 0.5 cm' of the medium below the soil sample so that the soil was in contact only with the vapor of the medium; and

c) By introducing 2 cm^3 of medium so that the soil was soaked with it.

Variation b) was inserted because it could be assumed that the life processes of eventual microorganisms might start even as a result of wetness (Oyama and Berdahl, 1977).

The changes in the composition of the gaseous phase above the soil sample were detected by gas chromatography, the instrument being capable of measuring H_2, Ne, N_2, O_2, Ar, or CO (the latter two could not be separated), NO, CH_4, Kr, CO_2, NO_2, and H_2S. The medium contained d- and l-amino acids, vitamins, other organic compounds, and inorganic salts. VL-1 investigated one original and one sterilized sample, whereas VL-2 analyzed one original and two sterilized samples. Sterilization was carried out at 145 °C for 3 h.

After wetting, N_2 and O_2 evolved from the original samples. As a result of the introduction of the medium, O_2 was consumed and CO_2 evolved. The only consequence of the addition of further portions of the medium was a small increase in CO_2 evolution. When studying sterilized samples, no change was observed in the amount of O_2 whereas the amount of CO_2 strongly decreased in the VL-2 experiment and slightly increased in the VL-1 measurement.

As can be seen, these measurements provided very contradictory results from the biological point of view. It soon turned out that the O_2 consumption after the addition of the medium corresponded to the oxygen uptake of the ascorbic acid in the soil. The other changes could not be explained by biological processes either. The increase in O_2 as a result of wetting was ascribed to the decomposition of peroxides in the soil, the increase in N_2 to desorption processes. The slow CO_2 evolution after the addition of the medium was explained by the oxidation of the organic components of the medium by the γ-Fe_2O_3 contained in the Martian soil.

In the third set of experiments [label release (LR) experiments], the phenomenon studied was whether carbon-containing gases (CO_2, CO) are liberated from the solutions of very simple organic compounds, on the addition of Martian soil. These compounds are common under extraterrestrial conditions on the one hand, and on the other hand, can be assimilated by terrestrial living beings. The experiment was very simple: 0.5 cm^3 of sieved Martian soil was introduced into the experimental cell, then the radioactivity in the atmosphere above it was measured. On average this was 490 counts/min. Then 0.115 cm^3 of a medium containing formiate, glycolate, glycine, d- and l-alanine, and d- and l-lactate in an aqueous solution was added; the individual concentrations of the components being 2.5×10^{-4} M, every carbon atom of each component of the medium was labeled. After the injection of the medium, the change in the activity of the atmosphere in the cell could be measured for as long as needed (in general for 6–7 days). The medium could be injected repeatedly (Lewin and Stroat, 1976).

Both VL-1 and VL-2 carried out several experiments with both active and sterilized soils. The results of all of these experiments were positive; moreover, they agreed very well in both their course and value, regardless of the site and time of the experiments. Therefore it is worthwhile to examine these experiments in more detail, and in the sequence of events.

The first experiment of VL-1 was made with active (not sterilized) Martian soil. After the addition of the medium, the activity of the atmosphere in the cell started to increase very strongly; a 15-fold increase of the initial 500 counts/min was detected on the first day. Then the increase slowed down and by day 6–7 it approached asymptotically an upper limit. This result exceeded every expectation (Fig. 10.4, curve 1). On seeing this result, the NASA spokesmen said that if these changes were really caused by Martian microorganisms, they were more active than terrestrial microorganisms.

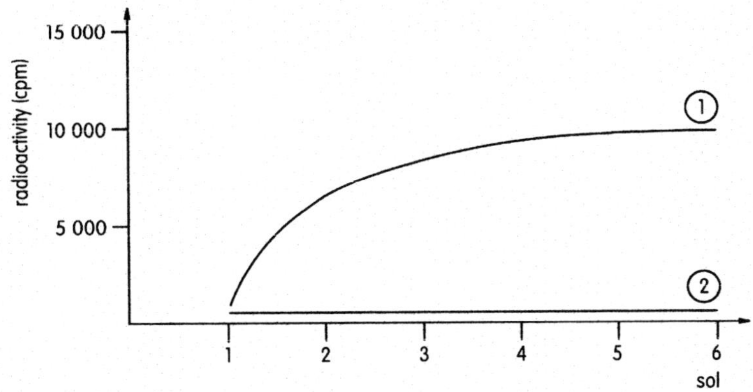

Fig. 10.4. One of the biological programs (LR) of the landing unit of the Vikings results in very positive answers. Curve 1: liberation of CO_2, from the medium exposed the effect of original Mars soil; curve 2: the same as for curve 1, but with previously sterilized Martian soil.

The next experiment was carried out with sterilized Martian soil (160 °C for 3 h). No change in the activity was observed at all (Fig. 10.4, curve 2). This fact seemed to support a hypothesis of the biological origin of the changes.

However, a more detailed analysis of the curve raised two problems. Biologists expected the initial section of the curve to be exponential and the end to be asymptotic. However, the most careful and thorough study of the curve could not find an exponential region, not even at its very beginning, whereas its end section was too short to establish an unambiguous asymptotic nature. Further lengthier experiments, among which was one lasting for 90 Martian days (sol), showed that the curve was not of an asymptotic character; rather it showed a very slow, but continuous increase. Upon repeating the experiments first by VL-1 and then by VL-2 the results obtained were always essentially the same (Lewin and Stroat, 1977).

However, in the meantime, it turned out from other experiments that the Martian soil was of a strongly oxidizing nature, contrary to expectations. Considering the relatively high ozone content in the atmosphere and/or the eventual peroxide content in the soil, it seemed more and more probable that the changes originated from the chemical oxidation of the components of the medium. Chemists showed, at the same time, that peroxides can also decompose as a result of heat treatment of above 100 °C, thus the negative result obtained from the sterilized soil sample was not necessarily a proof for the biological origin of the changes.

The experiments were then repeated at 46 and 51 °C inner temperatures and it became clear that the compound causing the activity is destroyed at temperatures of around 50 °C.

The very positive results of the LR experiments are ascribed usually to chemical reasons, mainly to the oxidation of the formiate contained in the medium, as the total activity produced corresponds to the activity originating from the complete oxidation of formiate and the kinetic course of the curve also supports this assumption. Oxidation may be performed by metal peroxides contained, or hydrogen peroxide formed, in the soil.

In spite of their excellent design and performance, the Viking spacecrafts did not succeed in finding any life processes on Mars. Therefore it seems certain that there are no living systems on Mars.

These experiments have been discussed in such detail not only because they are practically the only exobiological experiments that can be drawn from the literature. First of all, uncertainty was the characteristic prevailing when decisions had to be made concerning how is it possible to detect the presence of living systems in general, regardless of the type of their metabolism. For this, there was no method thought of beforehand, though two criteria were tacitly applied when analyzing the curves obtained. One of them is the expected exponential character at the initial section of the curves of the accumulation of metabolic products, the other one is the asymptotic nature of the curves.

Concerning the exponential character it should be noted that a) it is characteristic not only for living systems but for all self-supporting autocatalytic systems; and b) it is not necessarily characteristic for living beings. Living systems with program-controlled growth can show changes in the intensity of their metabolism that is completely different from the exponential, despite the autocatalytic nature of their metabolic network. It is readily possible, e.g., that for eventual Martian living beings permanent conditions for their growth and proliferation are provided only in well-defined astronomical situations; thus autocatalytic processes for ensuring their survival can be started only under these circumstances; in other conditions only a linear formation of nutrients is possible. Similar season- or weather-dependent phenomena are also well known on Earth.

Neither is the asymptotic nature of the metabolic intensity curve a characteristic of living systems only; it originates from the total consumption of nutrients and thus is characteristic for any chemical reaction, if the raw materials are not continuously supplied. LR experiments would supposedly also result in such a curve if formiate were the only oxidizable component in the medium: the slow linear increase is probably the result of the slow oxidation of the other components of the medium.

The Viking spacecrafts landed in desert areas, thus the Martian soil studied was practically dry drift sand. Comparative experiments should have been carried out at least afterwards with terrestrial desert drift sand. It is possible that the experiments – including ones trying to detect organic material – would have provided negative results even in this case.

Consequently, the Viking experiments cannot be accepted as reliable concerning living systems on Mars. However, we do not state that there is some form of life present on Mars, only that this cannot be excluded on the basis of our present knowledge. In addition, it is supposed that in the early periods of Mars it could have had a different, and for life more favorable, climate, and radar measurements indicate that at some regions of the planet liquid water can even now be found several centimeters below the surface.

Galilean moons

The four largest moons of Jupiter – Io, Europa, Ganymede, and Callisto – are called the Galilean moons because they were discovered by Galileo. They were studied thoroughly by the Voyager spacecrafts and the data obtained were very surprising not only because of the sulfur geysers and sulfur seas of Io already mentioned, but also because of the discovery that the other three moons contain large amounts of water in the form of ice and supposedly also in the liquid state. The structures of these moons are shown in Fig. 10.5 (Beatty, 1979).

Io has already been dealt with in detail; therefore here data concerning the other three moons are summarized. According to the high albedo and to the spectral data of Europa and Ganymede (Clark and McCord, 1980), their surface is covered mainly by water ice. The reflectance spectrum of the water ice on Ganymede shows that 65% of the surface is covered by water hoarfrost. The same measurement for Callisto is 20–30% (Mandeville, Geake, and Dollfus, 1980).

Europa is covered almost totally by water ice. The Voyager images of the moon show a complicated system of lines on the surface that presumably represent cracks in the ice crust. According to several authors, the wide gaps may be caused by the evaporation of liquid water originating from the time when the gases broke to the surface. The relatively thin (8 km?) ice layer protects the ocean underneath having a depth of about 100 km (!) from the coldness of space and prevents the heat generated by tidal changes from escaping, thus preventing the ocean from total freezing (Cassen, Reynolds, and Peale, 1979). The surface of Europa – disregarding the cracks – seems to be surprisingly smooth and the large numbers of craters, which are characteristic of celestial bodies without an atmosphere, are totally missing. This suggests that in geologically recent times the ice crust of the moon melted and froze again.

The ice layer covering Europa has a high light reflection and a light color. According to our present knowledge, water and organic material appear at the same places and so it should be assumed that they were also present simultaneously at the development of Europa. However, according to experience from chemical evolution, this leads to the formation of a liquid phase containing large amounts of dark brown organic material. The disappearance of this dark

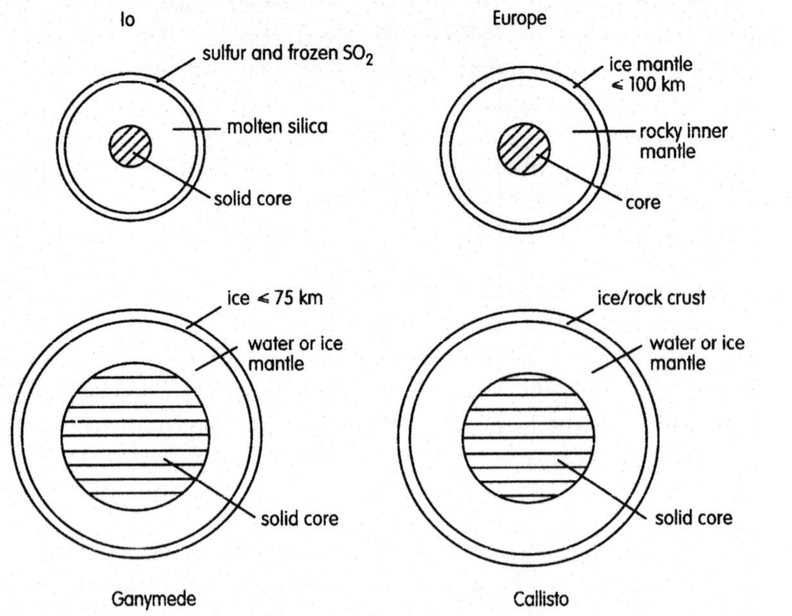

Fig. 10.5. The hypothetical structures of the so-called Galilean moons of Jupiter.

coloration can be explained, if biogenesis also occurred on Europa in early times and the living systems purified the water by collecting the organic material into themselves as had happened on Earth. Thus it is not impossible that in the initial period of its development, life also appeared on Europa.

Two researchers at NASA, E. Reynolds and S. Squyers, even supposed at the 1982 conference of the American Geophysical Union in San Francisco that there may be living systems present on Europa even now, because in the cracks of 1–70 km width and sometimes of length exceeding 1000 km water may appear induced by the great effects caused by the tidal forces in which the function of living systems may be ensured by the photosynthesizing effect of the Sun. If, however, the living world in deep-sea thermal springs on Earth is considered, where the energy needed for life is provided by the compounds with high energy contents formed by the interaction of hot, high-pressure water and basalt, even the assumption of Reynolds and Squyers is not necessary: the internal heat production of Europa originating from tidal forces is large enough and thus it can be assumed that at the bottom of the oceans covering it Europa "black smoker"-type thermal springs can be found that make the function of a chemosynthetic living world possible even independently of solar energy.

If, besides Earth, there is life at all in our solar system, it is not impossible that it resides in the depths of the oceans of Europa.

The surface of Ganymede is made up partly of water ice and partly of mineral materials, rocks, as revealed by Voyager images and spectral data. It is probably covered with a 70- to 100-km-thick ice crust; below it water or an ice mantle is to be found above the solid core. Its surface is significantly darker than that of Europa and strongly cratered. The surface of Callisto is even darker and more

Attribute	**Europa**	**Ganymede**	**Callisto**
Orbit radius (R)	9.47	15.1	25.6
Radius (km)	1552	2560	2420
Surface gravitation	0.13	0.15	0.12
Atmosphere	no	no	no
Temperature (K)	90-125	90-125	90-125
Solar constant (W/m^2)	52	52	52
Radiation	strong	probable	?
H_2O content (%)	5-20	50	?
Surface H_2O (%)	95	25	8
Ice, sleet, hoarfrost (% of surface)	100	65	20-30
Thickness of ice crust (km)			
Rocks in the surface	5-8(?)	75-100	150-200
	no	probable	certain

Table 10.4. Environmental conditions of the outer Galilean moons.

cratered; there is less ice and more rocks on its surface. The brownish color of the ice on Ganymede and Callisto suggests that their organic material that was formed during chemical evolution is at least partly contained in the ice. It cannot be excluded either that life appeared also in the waters of these moons in the early period of their development, but it is less probable than in the case of Europa.

However, it has to be emphasized that Ganymede and Callisto are the two celestial bodies where humans could set up a permanent settlement most easily (Almár and Gánti, 1981), as water and mineral materials are present for sure and organic material is supposedly also there on their surfaces. The radiation of the Sun provides enough energy for utilization (52 Wm^{-2}), the oxygen needed for respiration could be produced by decomposing water, and closed stations consisting of inflatable plastic constructions could be set up by sinking them into the ice by simple heating so that stability and protection against radiation would be performed by the ice (Gánti and Almár, 1982). The ecoconditions of the outer Galilean moons are summarized in Table 10.4.

Giant planets

The giant planets are Jupiter, Saturn, Uranus, and Neptune. They consist mainly of hydrogen and helium and do not have any solid surface. There are very little exobiological data about them. The mass of Jupiter is 318 times the mass of Earth, and Saturn's mass is 95 times that of Earth. Their mean density is 1.33 and 0.69, respectively; i.e., the density of Saturn is lower than that of water. Their atmospheres consist mainly of H_2 and He; CH_4 and NH_3 are present in small amounts, and ethane, acetylene, and PH_3 in traces. There are very strong currents in the atmosphere of both planets; the stream structures once formed are quite stable. One of these structures is the Great Red Spot, which has existed for at least 300 years. It is assumed that in their atmosphere chemical evolution processes take place and that the characteristic red and brownish colorations originate from the organic material formed in these processes. If the hypothesis described in Chapter IX that biogenesis occurred in floating droplets of primordial atmosphere is correct, then it is possible that in the atmospheres of Jupiter and Saturn, not only the chemical evolution processes are taking place, but prebiological evolution also occurs, and in the atmosphere of Jupiter and perhaps in that of Saturn primitive living beings exist. The Galileo program, however, in the frame of which a landing unit is planned for them, does not contain experiments of this nature.

XI. Outlines of a Quantitative Biology

Introduction

Recognition processes in biology started "from the top," i.e., its basic principles and concepts were set up on the basis of studying the most developed and thus most complex living beings. Biology had already been an advanced experimental science when studies on simplest living systems began, and possibilities for a detailed investigation of simplest organism have only become a reality in the last century. This direction of acquiring knowledge was obviously a necessity, but it resulted in contradictions, mainly concerning the fundamentals and the attitude of biology. For example, a taxonomic reform is going on at present for the system of the living world, and it is also a current task to find a satisfactory borderline between living and nonliving things.

A good example for this approach from more complex to less complex systems is the development of genetics. In the beginning, the mechanism of heredity was only imaginable on the basis of preformation: it was supposed that in the egg a whole living being was present on a smaller scale, otherwise the similarity in the parent–descendant line could not be imagined, even though this belief was in contradiction with the knowledge of that time. With the development of cytology it was then unambiguously proved that no preformed little living beings: in the ovum there is no small preformed living being present, but there is a nucleus containing chromosomes, and heredity is bound to the material of these chromosomes. Although this fact could be proved by diverse and well-reproducible experiments, the *mechanism* of heredity was thereby placed back into the realm of inexplicable myths, as it could not even be imagined how the property carrying hereditary information could be effectuated. This situation was not changed even when it turned out that the individual properties could be inherited independently of each other and thus specific material carried, genes, could be assigned to every property; neither when Avery showed that the material of genes was a chemically well-defined, uniform compound: deoxyribonucleic acid.

This myth was first broken by Schrödinger who pointed out that if the structure of the material carrying heredity is that of a "non-periodic crystal," then it is theoretically capable of storing information, even an unbelievable amount of information (Schrödinger, 1944). In 1953, the Watson—Crick model of DNA transformed this theoretical possibility into reality, and research work in the following two decades proved that heredity in the living world is, in fact, realized according to both the Schrödinger principle and the Watson—Crick model.

However, the reality behind the simple model proved to be extremely complicated, as biology revealed necessarily – and as a single possibility – the "high technology" of a 4 billion-year evolution. The complexity of the mechanism of transcription, translation, reparation, etc. now made the origin of heredity a myth: at present the biological attitude runs again into the usual obstacle: upon learning the most complex, the available body of scientific information itself starts to hinder the deeper elucidation of simple fundamentals, as already mentioned in connection with the origin of life.

Chemoton theory makes possible an opposite pathway. Starting at the simplest system, it goes towards acquiring qualitative and quantitative knowledge about more and more complex systems building them up gradually, as do geometry, mathematics, and organic chemistry. We have attempted to do just this from chapter to chapter starting at chemical reactions and designing from them the "constituents" of fluid chemical automata, then constructing different simple "fluid automata," later on more complicated self-producing, self-reproducing, and finally, proliferating systems. By constructing proliferating systems, we reached the minimal living systems, and by developing these further step by step we arrived at prokaryotic systems containing DNA genes and protein enzymes. It seems that this pathway follows the main line of the evolution of matter.

With systems at the prokaryotic level we undoubtedly entered the realms of biology. The author is convinced that this pathway may also be followed further within the realms of biology. However, at present it is only a belief: no one has yet completely followed this pathway to the end and neither could anybody have done so until now: the chemoton theory has come a long way in a short time from chemical reactions to prokaryotic systems, even though it did so at some places with omissions and in a sketchy way. Nevertheless, this belief of the author is not mere blind faith: it is supported by the results of investigations in different directions. Some of these will be shown in this chapter by also showing some advantages of this method, e.g., that this way the thought-channeling effect of the available knowledge hinders less the discovery of the unknown, unusual possibilities or that phenomena, which seem to be miracles or blind change from the opposite direction, this way become necessary and plausible events. At first some basic considerations of genetic nature will be shown.

Some Genetic Considerations

Self-(re)production without genetic subsystem ("technological recipe")

Self-production (self-reproduction) is not an exclusive property of living systems. When J. von Neumann elaborated the theory of self-reproducing automata, he thought that they should contain technological instructions for their operation and self-reproduction that should also be reproduced during self-reproduction (Neumann, 1948, 1966). Living systems, as has been discussed in detail in Chapter VI, without exception contain a genetic subsystem corresponding to the technological instructions of Neumann.

Autocatalytic reaction networks are self-reproducing networks characterized by the following general equation with the consideration of Eqs. [2.119], [2.120], and [9.1]:

$$\overline{A} + \left(2^{u} - 1\right)\overline{X} \xrightarrow{\textcircled{1}} 2^{u}\overline{A} + \left(2^{u} - 1\right)\overline{Y} \qquad [11.1]$$

Autocatalytic systems, though they are self-reproducing systems, do not contain a "technological recipe" or genetic subsystem.

"Proliferating microspheres" discussed in detail in Chapter IV are not only self-producing, but also proliferating systems. The construction principle of Neumann should be by all means applicable to them, if the general precondition of proliferation were really the presence of a "technological recipe." If the environmental conditions are not changed, microspheres produce identical descendants, because one of their subsystems, the autocatalytic reaction network, which produces the membrane-forming molecules, cannot change in a hereditary manner, and it only has the raw materials produced by itself for building the membrane. Thus the proliferating microsphere cannot undergo hereditary changes with respect to the nature of its material. On the other hand, its size and quantitative relations are determined by osmotic conditions, thus a variation in the osmotic conditions results in the appearance of descendants with different sizes. Changes in other thermodynamic parameters cause a difference in the ratio of the amounts of internal materials with respect to each other.

However, all of these changes are not hereditary but reversible, i.e., upon restoring the original conditions, proliferating microspheres identical to the original ones appear, independently of which generation the restoration occurs in and also of the changes that have eventually occurred during these generations. Thus these systems have the property of heredity in its narrower sense, but they are not capable of hereditary changes. They have no "technological recipe."

Template polymerization systems are macromolecular systems. In their self-reproduction the "parent" and "descendant" molecules should not only be identical chemically, but also in their macromolecular properties. It is a macromolecular property (the primary structure) making it possible that at this level of organization a qualitatively new property, the ability for the storage (and reproduction) of coded information, could appear. This means that monomers constituting the polymer may be utilized as signals; in homopolymers the amount of signals, whereas in co-polymers their amounts, their ratio, and their sequence may carry coded information. The latter is especially important, because it ensures an inexhaustible capacity for information storage.

The reproduction of macromolecules is determined not by thermodynamic parameters alone, but combined with the properties of the template molecule. Thus in the reproduction of macromolecules – differently than in the former systems – hereditary changes appear as a result of replication defects in a way considerably independent of the environmental conditions. For example, in homopolymer templates

$$pV_n + (n+1)V \xrightarrow{\textcircled{1}} pV_n + pV_{n+1} \qquad [11.2]$$

or

$$pV_n + (n-1)V \xrightarrow{\textcircled{1}} pV_n + pV_{n-1} \qquad [11.3]$$

types of defects may cause a hereditary shortening or lengthening of the reproducing macromolecule. In addition to these defects, in copolymers another type of defect, monomer exchange, may also occur, which can be represented as follows:

$$pV_nZ_m + (n-1)V + (m+1)Z \xrightarrow{\text{①}} pV_nZ_m + pV_{n-1}Z_{m+1} \quad [11.4]$$

This type of defect may cause hereditary changes in the macromolecule by leaving the length of the molecule unchanged.

It is easily to see that a unidirectional repetition of the event n times would result in a pure pZ polymer with a length of n+m:

$$pV_{n-1}Z_{m+1} + (n-2)V + (m+2)Z \xrightarrow{\text{①}}$$
$$pV_{n-1}Z_{m+1} + pV_{n-2}Z_{m+2} \quad [11.5]$$

$$pV_{n-2}Z_{m+2} + (n-3)V + (m+3)Z \xrightarrow{\text{①}}$$
$$pV_{n-2}Z_{m+2} + pV_{n-3}Z_{m+3} \quad [11.6]$$

etc

$$pV_{n-(n-1)}Z_{m+(n-1)} + (n+m)Z \xrightarrow{\text{①}}$$
$$pV_{n-(n-1)}Z_{m+(n-1)} + pZ_{n+m} \quad [11.7]$$

Exchange may, of course, take place also in the opposite direction:

$$pV_nZ_m + (n+1)V + (n-1)Z \xrightarrow{\text{①}} pV_nZ_m + pV_{n+1}Z_{m-1} \quad [11.8]$$

the repetition of which leads to the formation of a pure pV macromolecule with the same, n+m, length.

Finally, hereditary changes may also occur with unchanged length and composition of the template co-polymer, if the sequence of the constituting monomers changes. However, this change is not reflected in the stoichiometric equation, for a variation in the sequence does not cause any change in the mass balance. From this fact, in turn, a very important conclusion follows: Changes corresponding to Eqs. [11.2] - [11.8], concerning the information content of the template molecule, may be read in a purely chemical way (e.g., through the amounts of the raw materials used up in the template process), whereas changes in the sequence may be decoded only by specific sequence-reading mechanism.

Thus although self-reproducing macromolecules transfer their properties to their descendants in a hereditary way, the properties of the j-th generation may be significantly different from those of the original macromolecule. Therefore the properties of individuals of the j-th generation – in contrast to autocatalytic networks and self-producing microspheres – are determined not only by thermodynamic parameters, but also by the random hereditary changes that have occurred in their ancestors during generations. Thus in the self-reproducing macromolecules evolutional historicity already appears as a property-determining factor.

In spite of all this, template molecules do not contain a genetic subsystem or technological instructions: they determine themselves in their whole. They do not contain "coded" information, but serve as models for the formation of similar structures. However, they are capable of playing the role of an informational subsystem (technological recipe, genetic material) at a higher level of organizational hierarchy.

Reproduction by a "technological recipe"

In simple chemotons there is no reading apparatus capable of decoding the monomer sequence of the templates. However, changes in the template corresponding to Eqs. [11.2] – [11.4] and [11.7] can be appreciated by the chemical apparatus of chemotons. Such types of changes lead to hereditary changes not only in the template but also in the whole chemoton. For example, if a replication defect corresponding to Eq. [11.2] occurs in the template reaction, then the descendant containing pV_{n+1} and its descendants have information-carrying macromolecules of the site pV_{n+1} on one hand, and on the other hand each of them produces one more A and T and consumes one more V than the initial chemoton. Thus a replication defect of the type

$$pV_n \longrightarrow pV_{n+1} \qquad [11.9]$$

results in template polymerization in a hereditary change represented by

$$\left[\; n\overline{A} + pV_n + \boxed{T_n} \;\right] \longrightarrow \left[(n+1)\overline{A} + pV_{n+1} + \boxed{T_{n+1}} \;\right] \qquad [11.10]$$

in the chemoton generations, i.e., the change in the template has a hereditary change in the whole chemoton as a consequence. Thus pV_n is, in fact, a genetic subsystem of the chemoton, for it carries information for the whole of the chemoton in a "coded" form. The same naturally holds for changes of the Eq. [11.3] type.

Let us examine the effect of changes of the Eqs. [11.4] and [11.7] types by the following thought experiment. Let two types of chemotons proliferate in the same nutrient medium, i.e., in addition to chemotons of the

$$\left[\; n\overline{A} + pV_n + \boxed{T_n} \;\right] \qquad [11.11]$$

type, let chemotons be present having the following type:

$$\left[\; m\overline{B} + pZ_m + \boxed{T_m} \;\right] \qquad [11.12]$$

where the cycle

$$\overline{B} \equiv \sum_{j=1}^{1} \beta_j B_j \qquad [11.13]$$

is similar to cycle $\overline{A}$, its components being, however, all different. β_j is the fictive mole fraction of cycle $\overline{B}$ for which it holds that

$$\sum_{j=1}^{1} \beta_j = 1 \qquad [11.14]$$

and that cycle $\overline{B}$ produces a monomer Z that is capable of polymerizing on template pZ_m, and also capable of forming a co-polymer of the type $pV_n\, Z_m$ on a suitable template.

Thus the equation describing chemotons of the [11.12] type is

$$\left[\; m\overline{B} + pZ_m + \boxed{T_m} \;\right] + m\overline{X}^B \xrightarrow{\;(1)\;} 2\left[\; m\overline{B} + pZ_m + \boxed{T_m} \;\right] + m\overline{Y}^B \qquad [11.15]$$

Let us suppose that one or both types of chemotons unite into a giant chemoton, similarly to the fusion of the protoplasm. This may occur spontaneously – because their membranes have a two-dimensional liquid nature and are chemically identical – but may also be promoted in different ways. As a result, a chemoton corresponding to the composition of

$$\left[\ n\overline{A} + m\overline{B} + pV_n + pZ_m + \boxed{T_{n+m}}\ \right] \qquad [11.16]$$

is produced. The templates can also be coupled according to the equation

$$pV_n + pZ_m \longrightarrow pV_nZ_m \qquad [11.17]$$

but because stoichiometry considers only the amounts of V and Z incorporated into the template, stoichiometrically we are allowed to characterize the chemoton obtained by such a fusion, instead of Eq. [11.16], by the following composition:

$$\left[\ n\overline{A} + m\overline{B} + pV_nZ_m + \boxed{T_{n+m}}\ \right] \qquad [11.18]$$

independently of whether the template consists of the mixture of homopolymers or of co-polymers. These chemotons correspond to the following equation:

$$\left[\ n\overline{A} + m\overline{B} + pV_nZ_m + \boxed{T_{n+m}}\ \right] + n\overline{X}^A + m\overline{X}^B \xrightarrow{\ ①\ }$$

$$2\left[n\overline{A} + m\overline{B} + pV_nZ_m + \boxed{T_{n+m}}\ \right] + n\overline{Y}^A + m\overline{Y}^B \qquad [11.19]$$

(Such a chemoton obviously needs different osmotic surroundings than the original individual chemotons.)

However, in [11.19]-type chemotons hereditary information is coded not only by the number but also by the nature of the monomers in the template. Namely, replication defects resulting in exchanges $V \rightarrow Z$ or $Z \rightarrow V$ may occur by keeping the number of monomers unchanged. Thus in the presence of replication defects of the type

$$pV_nZ_m \longrightarrow pV_{n-1}Z_{m+1} \qquad [11.20]$$

the chemoton suffers a transformation that is described as follows:

$$\left[\ n\overline{A} + m\overline{B} + pV_nZ_m + \boxed{T_{n+m}}\ \right] \xrightarrow{\ ①\ }$$

$$\left[\ (n-1)\overline{A} + (m+1)\overline{B} + pV_{n-1}Z_{m+1} + \boxed{T_{n+m}}\ \right] \qquad [11.21]$$

This means that its size remains unchanged, whereas the quantitative relations of its metabolism are shifted in a hereditary way. A unidirectional, consequent n-time repetition of the [11.20]-type defects obviously results in the appearance of a pure $\overline{B}$-type chemoton having the composition

$$\left[\ (n+m)\overline{B} + pZ_{n+m} + \boxed{T_{n+m}}\ \right] \qquad [11.22]$$

and of size n+m. An m-time repetition of replication defects $Z \rightarrow V$ leads, correspondingly, to a chemoton having the same n+m size, but of the pure $\overline{A}$-type with a composition

$$\left[\ (n+m)\overline{A} + pV_{n+m} + \boxed{T_{n+m}}\ \right] \qquad [11.23]$$

It is apparent that replication defects of a mutational character occurring in the material carrying hereditary information lead to hereditary changes in the chemoton in spite of the fact that in these simple chemotons there is still no molecular reading mechanism capable of reading the sequence of the signals beyond the replication process itself. This is possible only because the replication defects in the composition, occurring not in the sequence but in the signals are fed back through a stoichiometric connection to the metabolism of chemotons, and thus the information-carrying subsystem of chemotons in fact plays the role of a control program, technological instruction, i.e., that of genetic material.

However, in Chapter IX it was also shown that through the evolution of chemotons, the living world could really reach the phase where the initially unintelligible "sequence of letters" was transformed into a text carrying intelligible information, and the realization of the utilization of this information was also developed, first in a direct way (eRNAs) and later in an indirect manner (through the translation mechanisms).

Hereditary properties acquired by adaptation

According to the gene theory, hereditary properties may only be changed through a variation in genes. Certainly, it is true that for a long time, while genes were supposed to be localized only in the nucleus, experimental proof was accumulated for a "cytoplasmic" heredity. However, at the end of the 1960s and beginning of the 1970s it was unambiguously proved that certain cell organelles (mitochondria, chloroplasts) have their own DNAs – and thus their own genes – and it became obvious that observations concerning cytoplasmic succession were not contradictory to the gene theory. In what follows it will be shown that in addition to, and independently of, mechanisms revealed by the gene theory, in some specific cases there is a theoretical possibility for hereditary changes that take place by adaptation in the course of generations. This type of heredity is not bound to macromolecules (Gánti, 1974b, 1979b).

Similarly to the derivation of the chemoton characterized by Eq. [11.19], let us consider also in this case the results of a thought-experiment concerning the fusion of two different chemotons. However, here the two chemotons should differ only in their autocatalytic cycles; their templates and the material of their membranes should be identical. This means that a chemoton with the usual composition of

$$\left[\ n\overline{A} + pV_n + \boxed{T_n}\ \right] \qquad [11.24]$$

is fused with a chemoton of the composition

$$\left[\ m\overline{C} + pV_m + \boxed{T_m}\ \right] \qquad [11.25]$$

where

$$\overline{C} \equiv \sum_{i=1}^{1} \gamma_i C_i \qquad [11.26]$$

and γ_i means the fictive mole fraction of the components of cycle $\overline{C}$ for which it has to be true that

$$\sum_{i=1}^{1} \gamma_i = 1 \qquad [11.27]$$

It is thus apparent that $\overline{C}$ of the chemoton characterized by Eq. [11.25] produces also V' and T', just as does it cycle $\overline{A}$ of chemoton [11.24], but through a different chemical reaction network, not having components identical with those of cycle $\overline{A}$.

According to the rules of cycle stoichiometry, the operation of the unified system should be characterized by the following equation:

$$\left[n\overline{A} + m\overline{C} + pV_{n+m} + \boxed{T_{n+m}}\right] + n\overline{X}^A + m\overline{X}^C \xrightarrow{\;①\;}$$

$$2\left[n\overline{A} + m\overline{C} + pV_{n+m} + \boxed{T_{n+m}}\right] + n\overline{Y}^A + m\overline{Y}^C \qquad [11.28]$$

However, this equation represents only one of the possible stoichiometric compositions, because cycles $\overline{A}$ and $\overline{C}$ produce V', and template pV_{n+m} only regulate the total amount of the components of $\overline{A}$ and $\overline{C}$ in a hereditary manner, without defining their ratio.

Thus in the fused chemoton stoichiometric compositions of

$$[\ (n-q)\overline{A}+(m+q)\overline{C}+pV_{n+m}+\boxed{T_{n+m}}\ ] \qquad [11.29]$$

$$[\ (n+q)\overline{A}+(m-q)\overline{C}+pV_{n+m}+\boxed{T_{n+m}}\ ] \qquad [11.30]$$

may occur, where q is an arbitrary positive number for which it holds that

$$q \le n \qquad [11.31]$$

and

$$1 \le m \qquad [11.32]$$

Thus the ratio $\overline{A}/\overline{C}$ inside this chemoton is determined by the initial amounts of the components of cycles $\overline{A}$ and $\overline{C}$ and by the overall rate of the two cycles. These, in turn, depend on the initial state, on the concentrations of the components, raw materials, and by-products, and on the rate constants of the elementary reaction steps, but do not depend on the properties of the template pV_{n+m}. Because cycles $\overline{A}$ and $\overline{C}$ are both autocatalytic cycles, the amounts (and thus also the concentrations) of the components of the cycle that is faster under the given conditions, rapidly exceed the concentrations of the components of the slower cycle. This process is advanced further by the increase of the reaction rate of the components as a result of the increased concentration.

If this process continues for a certain number of generations, the ratio of the concentrations of the two cycles may be shifted to such an extent that in the course of division not even a single molecule is incorporated into one of the descendants from the components of the slower cycle. In this individual, the ability to operate the slower cycle is lost, except for the case where some of the components are also produced in another metabolic pathway, or where some of the components are introduced into this individual from outside.

One should notice that this is, in fact, an adaptation process. This may be proved as follows. The reaction rates of the cycles depend to a great extent on the concentrations of nutrients $\overline{X}^A$ and $\overline{X}^B$. If the concentrations of the components $\overline{X}^A$ are high relative to those of nutrient components $\overline{X}^B$, the chemoton is enriched in the components of cycle $\overline{A}$ thereby increasing the rate of its own metabolism, thus decreasing its own generation time. After a certain number of generations, chemotons containing only cycle $\overline{A}$ appear that have hereditarily lost their ability to consume nutrient $\overline{X}^B$ or to produce V' or T' by means of cycle $\overline{B}$. Thus a hereditary change takes place through adaptation. If nutrient components $\overline{X}^B$ are in excess, this process proceeds in the opposite direction: components of cycle $\overline{B}$ accumulate and cycle $\overline{A}$ disappears.

This adaptive, hereditary change is not bound to macromolecules, it takes place without any previous change in the genetic material, i.e., it is independent of the genetic information contained in the macromolecules of the system. This principle holds also for the metabolism of cells, though their genetic material may contain the enzymes necessary for the internal transformation of the hereditarily lost cycle (network). It may be supposed that in cell fusion experiments, similar selection processes may take place in the metabolism of the fused cells, followed by the elimination of the useless part of the genetic material.

This means, at the same time that in living organisms there are properties, though in a small number, that is not at all, or not exclusively, bound to the genetic material. Genetic material may operate only within the hereditary limits defined stoichiometrically by autocatalytic cycles. This fact manifests itself in the trivial experience that no living system can be developed from mere genetic material (nucleus, spermatozoid), and supposedly it is also for the same reason that animal cells cannot be induced solely by gene transplantation to perform photosynthesis.

General structure of the genetic material

Watson and Crick have constructed the model of the molecular structure of the genetic material not on the basis of theoretical considerations, but based on the chemical and physical properties of a single chemically well-defined compound, DNA. According to preliminary chemical investigations, the structure should be that of a fibrous polymer built of deoxyribonucleotide units for which Chargaff's rule should hold, i.e., the molar amount of adenine should be equal to that of thymine, and that of guanine equal to that of cytosine (Chargaff, 1955–

1960). Density and viscosity studies have indicated that in the DNA fibers, two, eventually three, strands should run parallel, whereas other physicochemical studies have suggested that the bases are oriented toward the inside of the molecule and the phosphates toward the outside. According to the X-ray diffraction experiments of Astbury and Bell (1938) a periodicity of 3.4 Å is to be found in the molecule, whereas the experiments of Furberg (1952) showed that in nucleotides the planes of deoxyribose and those of nucleotide bases are perpendicular to each other. All of these available data have already outlined the possible structure of DNA (Gánti, 1962, 1966). The model constructed by the consideration of these data (Watson and Crick, 1953a) also proved to be a suitable model for the material carrying genetic information, as a by-product of the results (Watson and Crick, 1953b). The fact that the structure and operation of RNA are similar to those of DNA follows from their similar chemical structure.

However, the most important properties of the molecular structure of the genetic material can also be derived independently of the concrete chemical structure, by means of general principles based on the chemoton theory. The task to be performed by the genetic material is the storage of hereditary information and the production of copies for the descendants. If molecules serve as signals for storing information, they have to be fixed in space so that their position relative to each other does not change with time. This requirement may be fulfilled either by coupling the signal molecules to each other (polymerization), or by binding them to a carrier fiber or surface. (In nature both methods of information storage are realized: in proteins the signal molecules, the amino acids, are bound directly together, whereas in immune processes or in spermatozoa the signals are bound onto membranes.) In nucleic acids, if purine and pyrimidine bases are regarded as signals, the sugar phosphate chain is the carrier (fiber); if, however, individual nucleotides are regarded as signals, the nucleic acids may be considered as signal polymers without a carrier.

The fixation of molecules storing information may be carried out theoretically in one, two, or three dimensions, i.e., the "aperiodical crystal" of Schrödinger may in theory be a fibrous polymer, a membrane, or a three-dimensional crystal. For the fixation of hereditary information, however, the three-dimensional crystal is not suitable, for the information carried cannot be read in a chemical way because of the spatial inaccessibility and thus it can neither be reproduced chemically. Thus the remaining possibilities for signal fixation are the one- or two-dimensional arrangements. In nature, both can be found: the one-dimensional variant is widely used in the living world, whereas in the case of clay minerals the two-dimensional possibility is utilized (Weiss, 1981).

The simple chemoton model is capable of operating with signals fixed in both one and two dimensions; thus the simplest living beings in which the reading of the hereditary information occurs exclusively in a stoichiometric way may be imagined as having hereditary material of both the fibrous and the solid membrane structure, independently of whether information is carried in the amount of identical signals or in the ratio of the amounts of different signals (here the reader should be reminded that the sequence of signals cannot be read in a stoichiometric way). However, for both fibrous and membrane polymers the structural restrictions should hold as described for template polymers in Chapter II: template polymerization without enzymes is imaginable only if the bonds between the signals are stronger than those binding the signals to the template and the information carrier becomes stronger and stronger as the process proceeds and it is strongest at the end of the replication. From this it also follows that the most stable formation of the genetic material is the double helix structure for the one-dimensional case and the double membrane for the two-dimensional one. To start replication, in both cases a separation of the double structures is necessary; under nonenzymatic conditions this may take place also for membranes by the mechanism described in Chapter II.

The structure of the double-stranded fibrous polymer could theoretically be a ladder-like or a double helix structure. However, for the formation of the ladder-like structure the signals should be coupled by covalent bonds with a rigid 180° angle between them. This is such a strong restriction, which makes the formation of this structure highly improbable. Thus every double-stranded molecule is probably a double helix, independent of the actual chemical nature of the genetic material.

If the stable form of genetic material is a double structure (double helix, double membrane), and replication occurs on the surface of separated fibers (membranes), then one part of the newly formed structures is always the old one (serving as model) and the other part originates from the newly synthesized material (independently of whether the structure is of the fibrous or the membrane type). Thus the replication of the genetic material should always be of a semiconservative nature.

In our completely general (independent of actual chemical composition) consideration we thus arrived at the conclusion that the structure of the genetic material should be that of a double helix or membrane, replicating in a semiconservative way where the molecules constituting the individual fibers are bound by weak bonds (e.g., by hydrogen bonds). As is well known, genetic materials known so far in biology (DNA, RNA) completely meet all of these requirements.

In the next step, requirements concerning the number of the types of signals fixing information have to be considered. It is known that the living world generally and uniformly uses four different signals, the four nucleotide bases, for fixing information in genetic material. These bases in DNA are adenine, guanine, thymine, and cytosine, and in RNA they are adenine, guanine, uracil, and cytosine. However, it is also known that an arbitrary amount of information may also be stored by using only two signals. Why then does the living world use four signals after all?

The use of four signals may theoretically be an incidental event ("frozen accident"); a necessity originating from the concrete chemical structure, or a general law originating from the function and independent of chemical composition. In order to clarify this problem, we have to start from the general conditions for the pairing of the signal-carrying molecules necessary for replication by the template mechanism.

As has been shown in Chapter II, pairing could theoretically be homologous (in the case of homopolymers obviously only this), when V is coupled to V and Z to Z, or complementary, when V is coupled to Z and vice versa. In the first case the two halves of the double structure are identical (but of opposite directions), whereas in the second case the two parts complement each other (and are also of opposite directions), as illustrated in Fig. 2.18. Of course, the homopolymer carries information only by the amount of its uniform signals, whereas the copolymer may do this also by the sequence of its signals in addition to their amounts. However, for utilizing this type of information a suitable sequence reading (and utilizing) mechanism is necessary. As mentioned earlier, stoichiometrically only information carried by the amount of signals can be read.

For simplicity's sake, the general stoichiometric and kinetic equations of chemotons have been derived for homopolymer templates in which pairing are necessarily homologous. Upon studying the stoichiometry of mutational changes, homologous pairing was also implicitly supposed also in copolymers, i.e., V was always coupled to V and Z to Z. Variations characterized by Eq. [11.4] are possible only in this way in copolymers built of two different types of signals. Namely, if V is coupled to Z or vice versa, i.e., pairing is complementary, the amount of V must be identical to that of Z, which means that no changes of types Eqs. [11.4] – [11.7] and [11.8] are possible, i.e., "base-exchange" type mutations are stoichiometrically neutral.

It was discussed earlier in detail that, according to the chemoton theory, in the course of biogenesis, first such systems appeared in which the reading and utilization of genetic information occurred in a stoichiometric way, and then they underwent an evolutional optimization process through mutations. This process later on led to the appearance of enzymes, i.e., to the utilization of information stored in the sequence of signals. This process would be impossible in a co-polymer template containing only two signal types in the case of complementary pairing because in that case no stoichiometric change can take place in the information as the amount of molecules V and Z is 50–50% in molecule pV_nZ_m, i.e., n = m. This can be expressed so that in the case of complementary pairing, not the individual signals, but the signal pairs may be regarded as signals, and therefore, from the point of view of stoichiometry, the complementarily paired information carrier template molecule built of two monomers counts as a one-signal text.

However, if in the template molecule besides the V–Z signal pairs, another type of signal pair (let us denote it by W–S) is also present, i.e., the template molecule is built of four different signals, the double-stranded structure corresponds to that shown in Fig. 11.1, and mutations may appear that can be decoded in a stoichiometric manner as follows:

$$p(V,Z)_n(W,S)_m + (n-1)(V+Z) + (m+1)(W+S) \xrightarrow{①}$$

$$p(V,Z)_n(W,S)_m + p(V,Z)_{n-1}(W,S)_{m+1} \qquad [11.33]$$

Thus for prebiotic evolution to occur, genetic material had to be either of homologous pairing and built of a minimum of two types of signals, or of complementary pairing and consisting of a minimum of four different signals (Gánti, 1974b). The genetic material of the terrestrial living world prefers complementary pairing and is built of four different signals. Thus if prebiotic evolution took place in the way derived earlier on the basis of chemoton theory, nucleic acids were built of four bases because this was the theoretically simplest way of storing genetic information. This shows that the conclusion following from chemoton theory, i.e., that at the beginning of prebiotic evolution informational operations took place in a stoichiometric way, is correct. In the case of non-stoichiometric informational operations, nucleic acids having only two types of signals would only be required for carrying genetic information.

Anything that was said about pairing is equally true for fibrous and membrane-like materials. As long as informational operations take place in a stoichiometric way, i.e., in the first period of prebiotic evolution, genetic material may equally be of the double strand or double membrane structure, in the case of homologous pairing with a minimum of four types of building units.

In the second period of prebiotic evolution, the process enabling the development of enzymes is loop formation. For this purpose, information carriers in the membrane form are not suitable. Thus prebiotic systems with membrane-structured genetic material could not go beyond the level of evolution ensured by stoichiometric informational operations, and could not enter the phase where information could be carried by the sequence of signals.

Hence it seems to be a general rule that the genetic material of a well-developed living world has to be a macromolecule with a double-helix structure and having template properties is built of a minimum of four different types of building units if pairing is complementary (Gánti, 1974b, 1979b).

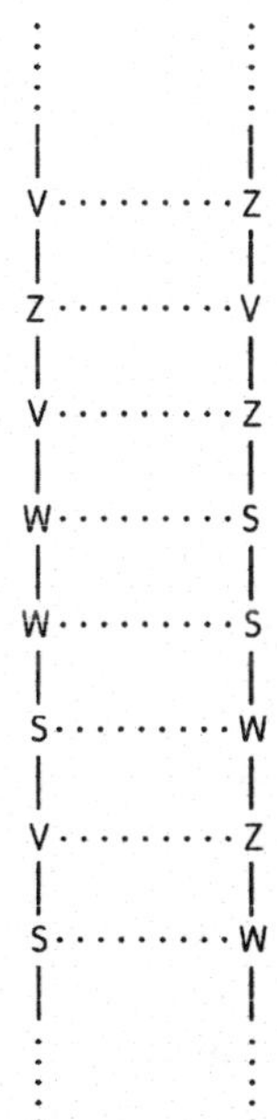

Fig. 11.1. A template polymer built of four types of building stones with complementary pairing and having a double-stranded structure.

Quantitative Treatment of a Prokaryotic Cell

Possibilities and limitations

In Chapter VIII it was shown that cycle stoichiometry and stoichiokinetics developed in the first volume are suitable not only for the quantitative characterization of the abstract chemoton model, but chemotons containing real compounds can also be accurately calculated by their application. As an example, the calculation of the stoichiometric changes between two divisions of the total chemical supersystem of a hypothetical prebiotic chemoton was presented. On the other hand, in Chapter IX it was demonstrated that prebiotic chemotons could possibly develop into a prokaryotic cell as a result of a long evolutional process, thereby resulting qualitatively the simplest biological units of today. The next step would be the quantitative description of these prokaryotic cells. In what follows, we make an attempt to provide a solution of this task.

The simplest abstract chemotons contain the appropriate self-producing cycle corresponding to the metabolic network, the membrane system, and the homopolymeric template macromolecules on which the informational operations are based. The example of the concrete prebiotic chemoton shown was more complex: its metabolic network consisted of several cycles and networks (formose network, acetylphosphate regeneration cycle, *de novo* acetylphosphate synthesis, nucleotide cycle) and of the linear reaction pathways connecting them and coupled to them, and the information-carrying macromolecule was also built of four different building units. In prokaryotes, metabolic networks become even more complicated. Nevertheless, this is not the essential difference between them, rather the appearance of macromolecular metabolism, i.e., that of protein and polymeric carbohydrate metabolic networks, and in connection with this the transplantation of the gene functions into DNA and the development of new RNA functions (messenger, transfer, ribosomal functions). Thus, if we want to discuss the prokaryotic cell in an exact, quantitative manner, these should also be built into our equations.

A quantitative description can be done in two possible ways: the algebraic (in our case, stoichiometric) way and the kinetic one. Attempts in biology and biochemistry have followed so far mainly the kinetic approach (see, e.g., Reiner, 1969; Oster, Perelson, and Katchalsky, 1971; Reich and Sel'kov, 1974; Rössler, 1974a, b; Heinrich, Rapoport, and Rapoport, 1977; Beretta et al., 1979). The reason for this is partly that enzymes cannot be included in stoichiometric equations and thus their quantitative treatment is possible only kinetically, and partly that in the related theoretical branches of science such as system theory, cybernetics, and irreversible thermodynamics, kinetic approaches also predominate.

Though there is no theoretical difficulty in the quantitative description of the behavior of a cell of any complexity, in practice it is hopeless and will remain hopeless for a long time to come. To do this an equation system consisting of about 10^4 partly nonlinear differential equations has to be solved, even in the case of the simplest cell. Even numerical solutions cannot be expected, not only because of the enormous computer capacity needed, but also because for this purpose the knowledge of the same number of reaction kinetic constants and concentration values would be necessary. It is no mere chance that researchers have been restricted so far to networks consisting of very few elementary steps or else they contain negligence (usually multiple negligence) that make the results unacceptable.

In contrast, stoichiometric treatment seems to be promising, and, within certain narrow limits, stoichiokinetics may even provide information about the time course of reactions. In the stoichiometric discussion of the prebiotic chemoton a network consisting of about 10^2 elementary reaction steps could be treated quantitatively. The fact that a chemical network contains 10^4 reaction steps instead of 10^2 does not entail any specific difficulty, all of the operations being here simple arithmetical ones. There is no need to know the kinetic constants and concentrations, only the chemical formulas of the components of the network should be known most of which are already known. If some unknown components are sporadically present in the network, their composition can even be determined by cycle stoichiometry applied in calculations.

Though the number of the elementary steps does not raise any difficulty, the complexity of the network does: at the time being, cycle stoichiometry is elaborated only for cycles and simpler networks in an exact way. Difficulties arise in networks containing more complicated branchings and connections. According to preliminary studies, these problems can be solved by the application of graph theory and matrices; however, the rules of their use are not as yet elaborated in general, and thus the metabolic network of a prokaryotic cell cannot presently be treated quantitatively in full. Nevertheless, we may follow the same way as in the derivation of the chemoton model: we construct a minimum model for the prokarytic cell and then derive its equation.

Stoichiometry of the minimum model of prokaryotes

As a first step, the minimum model of chemotons containing DNA genes but RNA enzymes will be constructed. As was derived earlier, it might be supposed that during evolution, first RNAs had evolved into enzymes, before proteins were synthesized. It was also shown that as a result of the template mechanism, the sequence of any replicas of eRNAs is necessarily complementary to the original RNA, i.e., if one of the strands of the double helix carries an enzyme function, the other one must be its gene. This hereditary mechanism allows for a very large variability thereby ensuring a very fast evolutional optimization. However, after this optimization process is over, this large variability is not advantageous any longer and in fact is rather unfavorable. According to our hypothesis, at that time DNA appeared and took over the role of genetic information storage from gRNAs.

The development of the DNA-based metabolism was rendered possible by the eRNAs ensuring the deoxyribonucleotide synthesis and the development of nucleic acid polymerase eRNAs. Namely, by means of the former, the raw materials needed for DNA synthesis could be produced, whereas the latter ensured the transcription of the genetic material already optimized at the RNA level into the DNA language and could also ensure that eRNA replicas could be made from the gene DNAs formed. The appearance of RNAs with polymerase functions (RNA polymerase RNA and DNA polymerase RNA) has other consequences as well. First it can significantly reduce the probability of replication defects, i.e., it contributes by itself to the increase in the stability of the optimized gene population. Another effect is that it can supersede the nucleotide concentration-dependent mechanism (v^*) of the initiation of replication described in Chapter II by substituting it for a more complex, enzymatic mechanism. Thus, in turn, the replication of DNA and RNA may become independent of each other, RNA synthesis on an RNA template, DNA replication on a DNA template as well as RNA synthesis on a DNA template become possible. The first process only plays a role in the transcription of optimized gRNA populations into DNA (therefore it is expedient for the corresponding eRNA to be eliminated in the course of evolution), whereas the other two processes are involved in the survival of chemoton populations with gDNA–eRNA combination, thus their presence is required until their function is taken over by the corresponding enzyme proteins in a subsequent phase of evolution.

Finally another important consequence is that the enzyme mechanism of the initiation of replication processes enables replication to start at lower v^*

threshold values that do not disturb the tertiary structure of eRNAs. Thus in the individual chemotons eRNAs may remain active during the whole generation cycle. This again means an evolutional advantage over those chemotons in which the eRNA is denatured during gene manipulations (or enzyme synthesis) and which therefore are in an inactive state.

The metabolic map of the minimum model of chemotons containing eRNAs but gDNA is illustrated in Fig. 11.2. Notation D refers to deoxyribonucleotides, pD_m is the single-stranded nucleic acid and the subscript m indicates that the total of the DNAs of the given chemoton contains m deoxynucleotides. pD_mD_m denotes the double-stranded structure. Ribonucleotides are denoted by V, RNAs by pV_n. DNA is synthesized on a DNA template, pV_n (i.e., eRNA) on the double-helix-structured DNA, thus the replication cycles of the two sequence-carrying polymers are coupled. V' is the common precursor of V-s and D-s, i.e., an angle-point of an OR-branching in the metabolic map. At the beginning we showed that an OR-branching in itself is undefined stoichiometrically. However, built into this metabolic network it becomes stoichiometrically definite, as the consumption of V and D is determined by the template process, i.e., the ratio of D and V consumed is: $D/V = m/n$ (here we neglect the production of D and V needed for reaching the reaction equilibrium, as this can also be calculated for any particular case). The meaning of P will be discussed in the following phase of developing our model.

It is apparent that differently from the minimum model of chemotons, four autocatalytic cycles are coupled here, not three. On the basis of the figure, we can write the cyclic equations of the individual subsystems, and by adding the linear synthesis pathway leading to membrane formation, we obtain the overall equation of the given chemoton.

$$pD_m + mD \xrightarrow{\textcircled{1}} 2pD_m + mR \qquad [11.34]$$

$$pD_mD_m + nV + pV_n \xrightarrow{\textcircled{1}} pD_mD_m + 2pV_n + nR \qquad [11.35]$$

$$\boxed{T_{m+n}} + (m+n)T \xrightarrow{\textcircled{1}} 2\ \boxed{T_{m+n}} \qquad [11.36]$$

$$(m+n)(T' + R) \longrightarrow (m+n)T \qquad [11.37]$$

$$(m+n)(A_1 + X) \xrightarrow{\textcircled{1}} (m+n)(A_1 + Y + V' + T' + P) \qquad [11.38]$$

$$(m+n)V' \longrightarrow mD + nV \qquad [11.39]$$

$$\left[(m+n)A_i + pD_m + pV_n + \boxed{T_{m+n}}\right] + (m+n)X \xrightarrow{\textcircled{1}}$$

$$2\left[(m+n)A_i + pD_m + pV_n + \boxed{T_{m+n}}\right] + (m+n)(Y + P) \qquad [11.40]$$

Thus Eq. [11.40] is the overall equation of a minimum model of a chemoton containing DNA as genetic material and RNAs as enzymes. The next step of evolution is the appearance of enzyme proteins. In our model, we naturally discuss only its stoichiometric aspects. Correspondingly, to the metabolic map shown in Fig. 11.2, a protein-synthesizing subsystem should be added. If this is done, our system is a minimum model including everything required of the simplest, presently existing prokaryotic cell: a genetic apparatus consisting of DNA, a translation apparatus based on RNA, a protein synthesizing system, a membrane system, and a metabolic network.

All of these will be discussed only from the stoichiometric point of view. Thus this minimum model is simultaneously also the minimum model of prokaryotes, and the equation describing its operation is the equation of a prokaryotic minimum model.

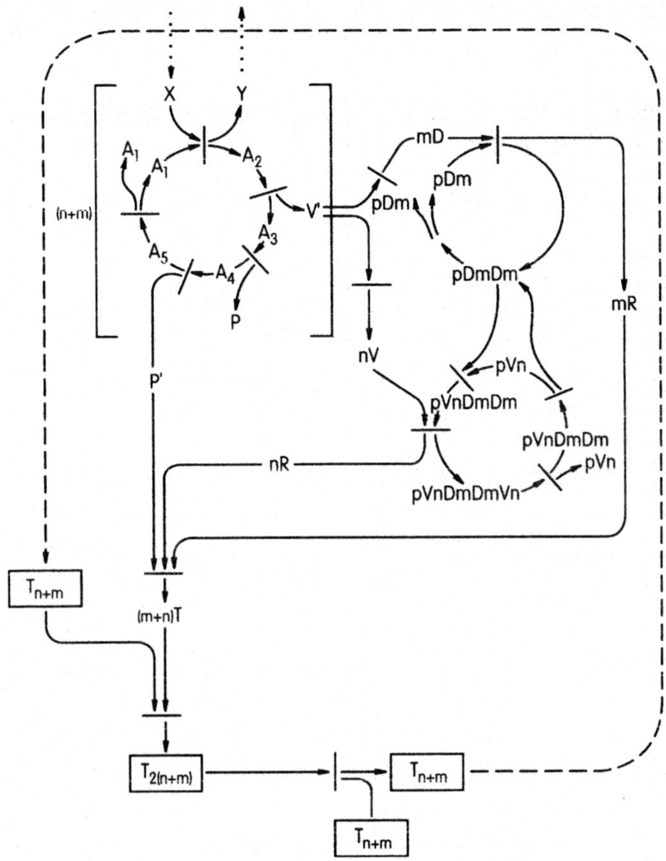

Fig. 11.2. The theoretical scheme of chemotons containing enzyme RNAs but DNA genes.

However, before performing the addition of the protein-synthesizing subsystem, we have to consider the stoichiometric role of enzymes in the metabolic network. The corresponding considerations have already been applied implicitly in the model characterized by Eq. [11.40], thus it is useful to examine the problem in this model. In Fig. 11.2, eRNAs appear only in a single subsystem in the autocatalytic cycle $pV_n \xrightarrow{①} 2pV_n$, i.e., in the process in which they are synthesized. Because we assumed that in this system the individual metabolic steps were catalyzed separately by specific eRNAs that we denoted by $pV_{n1}, pV_{n2}, \ldots, pV_{nm}$, and for which it should hold that

$$pV_n = \sum_{j=1}^{1} pV_{n_j} \qquad [11.41]$$

in a detailed metabolic map, every reaction step should be substituted for a cycle the cycle component of which is the pV_{nj} specific to the given reaction step. For example, the elementary steps of cycle $A_1 \xrightarrow{①} 2A_1$ characterized by Eq. [11.38] are:

$$A_1 + X + pV_{n_1} \longrightarrow A_2 + pV_{n_1} + Y \qquad [11.42]$$

$$A_2 + pV_{n_2} \longrightarrow A_3 + pV_{n_2} + V' \qquad [11.43]$$

$$A_3 + pV_{n_3} \longrightarrow A_4 + pV_{n_3} + P \qquad [11.44]$$

$$A_4 + pV_{n_4} \longrightarrow A_5 + pV_{n_4} + T' \qquad [11.45]$$

$$A_5 + pV_{n_5} \longrightarrow 2A_1 + pV_{n_5} + T' \qquad [11.46]$$

$$A_1 + \sum_{j=1}^{5} pV_{n_j} + X \xrightarrow{①} 2A_1 + \sum_{j=1}^{5} pV_{n_j} + Y + V' + P + T' \qquad [11.47]$$

Thus, in fact, instead of Eq. [11.38], Eq. [11.47] should be used in calculating overall equation [11.40]. Similarly, in Eqs. [11.34], [11.37], and [11.39] the specific pV_{nj}-s catalyzing the elementary reaction steps should also be included. In contrast, in Eq. [11.35] only a single pV_n plays the role of a catalyst, the DNA-dependent RNA polymerase (let us denote it by pV_{nk}), thus this equation is in its correct form:

$$pD_mD_m + pV_{n_k} + nV \xrightarrow{①} pD_mD_m + pV_{n_k} + pV_n + nR \qquad [11.48]$$

However, if the overall equation of the whole system is calculated by the summation of elementary equations also containing enzymes, because of the correlation formulated in Eq. [11.41] an equation identical to Eq. [11.40] is obtained. In formulating Eqs. [11.34] – [11.39], we have taken into consideration the stoichiometric role of enzymes not separately, in the processes catalyzed by them, but globally in their formation (i.e., in Eq. [11.35]). Our procedure for the calculation of the stoichiometric overall equation of systems containing protein enzymes will be similar.

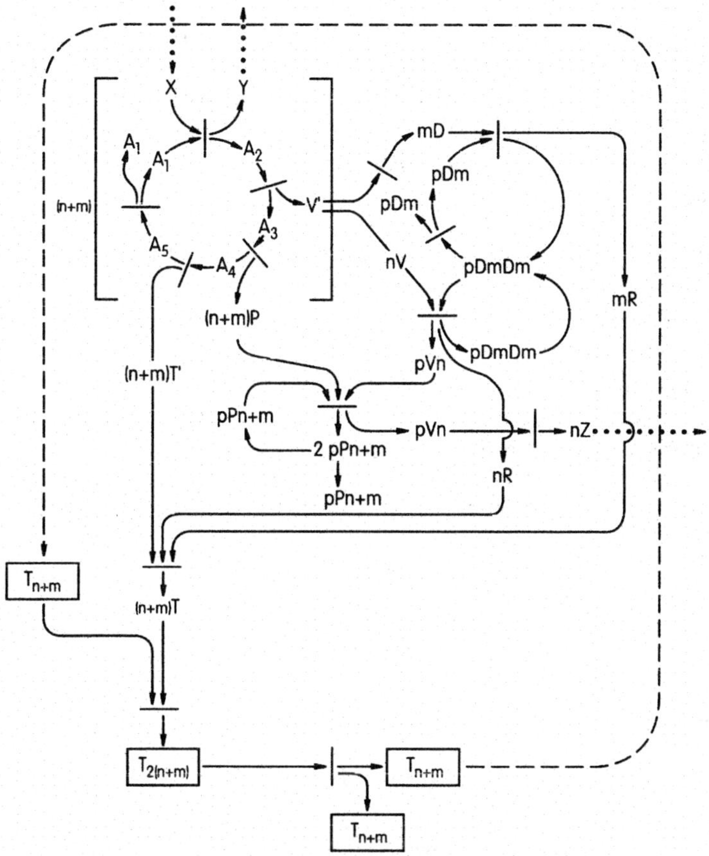

Fig. 11.3. Theoretical scheme of chemotons containing DNA genes, RNA synthesis, and protein enzymes. This model can also be considered the minimum model of prokaryotes.

By including the protein-synthesizing system, the functions of RNAs are taken over by protein enzymes also including the polymerase function, thus the autocatalytic role of the RNA subsystem also ceases to exist. At the same time, though the protein synthesis in itself is not autocatalytic, it is autocatalytic with respect to the whole system, as the operation of all protein enzymes is required for protein synthesis within the system, if only indirectly. If we denote the amino acids by P, then by analogy, the designation of proteins should be pP_n (similarly to pV_n and pD_m), and thus the overall equation for protein synthesis can be written as

$$pP_n + nP + pV_n \xrightarrow{\text{①}} 2pP_n + pV_n \qquad [11.49]$$

This indicates that protein synthesis is an autocatalytic process but does not show that this autocatalytic nature is true only as a part of the whole proliferating system and that it is realized only indirectly. If, however, we proceed in the derivation of the stoichiometric overall equation of the minimum prokaryotic system according to these principles, the individual consideration of the role of enzymes in the elementary reaction steps of the whole system becomes unnecessary.

Figure 11.3 shows the metabolic map of a prokaryotic minimum system produced according to the above considerations. In this map, the pV_n representing RNA, after performing its function, decomposes to Z, which leaves the system as a waste material. However, this is not necessarily so, the decomposition product may be repeatedly utilized, but this complicates the map. The stoichiometric overall equation of the system shown in the figure is obtained by the summation of the following equations:

$$pD_m + mD \xrightarrow{\text{①}} 2pD_m + mR \qquad [11.50]$$

$$pD_mD_m + nV \xrightarrow{\text{①}} pD_mD_m + pV_n + nR \qquad [11.51]$$

$$pV_n + (n+m)P + pP_{n+m} \xrightarrow{\text{①}} 2pP_{n+m} + pV_n \qquad [11.52]$$

$$\boxed{T_{m+n}} + (m+n)T \xrightarrow{\text{①}} 2\boxed{T_{m+n}} \qquad [11.53]$$

$$(m+n)(T' + R) \longrightarrow (m+n)T \qquad [11.54]$$

$$pV_n \longrightarrow nZ \qquad [11.55]$$

$$(m+n)(A_i + X) \xrightarrow{\text{①}} (m+n)(2A_i + Y + V' + T' + P) \qquad [11.56]$$

$$(m+n)V' \longrightarrow mD + nV \qquad [11.57]$$

$$\left[(m+n)A_i + pD_m + pP_{n+m} + \boxed{T_{m+n}}\right] + (m+n)X \xrightarrow{\text{①}}$$
$$2\left[(m+n)A_i + pD_m + pP_{n+m} + \boxed{T_{m+n}}\right] + (m+n)Y + nZ \qquad [11.58]$$

Equation [11.58] is the stoichiometric overall equation of a minimum prokaryotic system. An infinitely large number of more and more complicated variants of this model may be constructed. It can be assumed that in nature, in the initial period of evolution; those had been realized for which the materials and conditions required were present. In the course of an optimization process, the majority of them became extinct and only those survived that were of an optimal function for the terrestrial ecosystems. Even the number of these variants is enormously high.

The Stoichiometric Principles of Evolution

Csányi's general theory of evolution

It would perhaps be more correct to use the expression "biological evolution" instead of "evolution" here, as our subject in the last third of this book is also concerned with evolution (prebiotic evolution), and as the word "evolution" is also used in fields far from biology, such as chemical evolution, technical evolution, cultural evolution, etc. However, in what follows, the part of history will be discussed for which the word "evolution" has originally been used: the development of the present living world from the primordial cell (or cells). If, namely, the pathway from chemical reactions to the prokaryotic cell (or at least a possible variant of this way) has been possible to be followed by means of the chemoton theory, we now have to examine whether the principles and methods are capable of describing the way leading from prokaryotes to multicellular beings, i.e., the indisputably biological region. It should be emphasized that we do not attempt to substitute Darwin's theory or any neo-Darwinist theory with the chemoton theory, or even to modify them. The chemoton theory is not suitable for this purpose. However, as for a mechanical engineer, geometry provides the possibilities and limits for design of machines, after appropriate development of the cycle stoichiometry provides the chemical possibilities and limits for the construction of variants of the metabolic system. At the prokaryotic level, these chemical possibilities and limits are predominant; however, by the proceeding of evolution, more and more other determining factors also arise. These other factors are in the limelight of biology today, although e.g. chemotaxonomy is based just on the differences between metabolic networks. The importance of chemical stoichiometry even at the most complicated biological levels is supported e.g. by metabolic diseases, the nutrient problems in plant cultivation, or the environmental problems, the majority of them manifest themselves just in the stoichiometric shifts in the chemical networks of ecosystems.

It should also be mentioned that the stoichiometry of these fields is not at all elaborated contrary to the systems discussed so far. Thus we can restrict our studies only to the principles enabling or restricting the development of the living world from the stoichiometric viewpoint.

Biological evolution involves at least four levels of the organizational hierarchy built upon each other: the levels of prokaryotes, eukaryotes, multicellular beings, and systems organized at a supraindividual level. In our study it would be useful to choose a general evolution theory for discussing the stoichiometric regularities that is applicable at all four levels. For this purpose the most suitable theory seems to be that of Csányi (1978, 1980, 1981, 1982, and 1988). Although the basic model of this general evolution theory, the so-called autogenetic model (Csányi and Kampis, 1984, 1985), and the corresponding definitions are not yet fully developed, Csányi successfully applies his basic hypothesis for the different levels of evolution (prokaryotic, eukaryotic, multicellular levels, ecosystems, neural evolution, technical evolution) in his most recent work (Csányi, 1988). These applications allow the conclusion that the autogenetic model is basically correct, and the faults and inconsequence in its present form can be eliminated by a further development of the model. In what follows, the essential features of the autogenetic model will be outlined, but only with respect to their applicability to our system.

It is well known that each system standing at a level of organizational hierarchy may serve as a system-forming element at the next level of organization (e.g., the atom as a system is an element of the next organizational level of hierarchy, namely, at the molecular level; or the cell as a system is a system-forming element at the level of multicellular systems). Csányi calls these system-forming elements *components.*

The assembly of components at a given level of organizational hierarchy may be: (1) unorganized assembly, (2) an organized assembly in which the type of organization important from the point of view studied is not realized, and (3) an assembly in which the organization according to the viewpoint studied is (also) found. For example, in the evolution of multicellular beings, in principle, the assembly of unorganized eukaryotic cells may be assemblies of the first or second type, but an assembly of eukaryotic cells having some, e.g., ecological types, if there are no multicellular beings among them. Thus the general formulation of genesis with respect to a given level of organizational hierarchy is, how an assembly of type 1 and 2 transforms into an assembly of type 3. The assemblies of types 1 and 2 are called by Csányi zero-systems (0-system).

Between the components of an assembly different interactions operate. Csányi's very important recognition is that part of these interactions become functions with respect to the formation, operation, and organization of the system in question, i.e., they transform into interactions having definite roles in the formation and operation of the system. For example, the fact that a magnet attracts a piece of iron is an interaction, but in an electric bell it is a function bringing about mechanical motion required for sound generation. However, not only an interaction, but also a property may be a function: such as conductivity in our example.

The systems discussed are proliferating (in Csányi's terminology: reproductive) systems. A prerequisite for proliferation of a system is the reproduction of its components, i.e., the formation of newer and newer pieces of components. Interactions playing a role in the organization of the system influence directly or indirectly, the probability of the formation of the other components. Therefore, interactions by which the components are capable of influencing the formation probability of each other (and thus the reproduction of the system) will be called *functions*, or more accurately, *reproductive functions*.

After Gray (1975), Csányi also uses the concept of system precursor. According to his definition, the autogenetic system precursor (AGSP) is a minimum component network yet capable of replication satisfying the following requirements:

a. It contains at least one component-producing cycle that is closed with respect to replicative information.
b. At least one component of this cycle can be excited by the energy current passing through the system.

The process of evolution may be described by means of the autogenetic model in the following (to a great extent simplified) way: At a given level of organizational hierarchy there is an assembly of the multitude and variety of components (0-system). There are different interactions operating between the components and types of components. If the system is suitable for the formation of a network of functions satisfying AGSP requirements as a result of the organization of interactions (i.e., the assembly is a 0-system), the specific AGSP develops in it, which results in the organization of the continuously developing complexity in the 0-system (autogenesis). Thus AGSPs are functionally (but not necessarily structurally) organized replicative systems (cycles). There is some collaboration and *functional* differentiation between these replicative cycles, "communities" of jointly replicating components, thus supersystems, supercycles appears. The process is closed by the spatial separation and compartmentalization of the supersystem.

If compartments are also replicated, we arrive at reproductive supercycles or supersystems, i.e., at spatially proliferating systems. (Attention should be called here to the difference between Csányi's interpretation of replication and reproduction and that used in this book. However, this will probably not interfere with the comprehension of the essential concept.) If compartmentalized supersystems are proliferating, a multitude, an assembly of them will develop. In the case of the formation of different such systems (which may be the result of both *de novo* formation and non-identical replication), an assembly of their multitude and variety may develop with specific interactions between them. The different interactions make possible their organization into a functional system, thus the supersystems become components (at a higher level of organizational hierarchy) and they acquire functions. Hence at this higher level of organizational hierarchy the assemblies become 0-systems and in them the formation of AGSPs corresponding to this level starts, followed by the development of supercycles, compartmentalization, etc.

Obviously, by further repetitions of this process a series of organizational levels can develop in the organizational hierarchy of reproductive systems, provided the conditions are appropriate. Moreover, because AGSPs may be formed by different principles, parallel evolutional processes can take place at the same level, for which good examples are the different evolutional process of the supraindividual level (cultural evolution, technical evolution, political evolution, etc.). It should be emphasized again that the picture outlined here is a one-sidedly simplified version of Csányi's general evolution theory. Csányi's theory also uses concepts that are not even mentioned here (e.g., replicative information, functional information, parametric information, complexity), however, they are dispensable from the viewpoint of the stoichiometric study of evolution.

Hence it is apparent that the picture of prebiotic evolution drawn by means of the chemoton theory is in full accordance with the autogenetic model. The components of the 0-system are the organic compounds formed in chemical evolution; part of their chemical interactions makes the formation of replicative cycles possible, whereby these chemical interactions become functions at a metabolic character. This is how the prebiotic autogenetic system precursors come into existence: the formose cycle, template polymerization, and the two-dimensional liquid membranes. Stoichiometric interlockings are established between them and thus chemotons (supercycles) are formed. During their proliferation, chemotons first replicate with a high mutation probability, therefore a great variety of chemotons develop, becoming more and more complex in the process. This process, however, decreases the mutation probability (DNA genes, formation of protein enzymes, enzymatic template polymerization). The chemotons proceed from the phase of non-identical replication to the phase of identical replication, reaching meanwhile the prokaryotic evolutional level, as has already been described in detail in the chapter on prebiotic evolution.

Stoichiometric principles of the evolution of prokaryotes

Csányi's autogenetic model fully corresponds to the process of prebiotic evolution outlined on the basis of the chemoton theory. The question is whether the chemoton theory can also be applied within the frame of the autogenetic model at the next level of organizational hierarchy where prokaryotes become components. In order to decide this, the following questions should be answered:

1. Can an assembly of the multitude and variety of prokaryotes develop?
2. Are there any interactions between prokaryotes?
3. Are there interactions that may be ordered into cyclic chains of events, i.e., into replicative cycles, which means that functions appear between prokaryotes?
4. Can AGSPs organized from prokaryotes as components appear?
5. Does functional differentiation come about?
6. Does compartmentalization take place?
7. Can all this be described and calculated quantitatively by means of the chemoton theory?

The condition that the assembly of a multitude of prokaryotes develops under suitable circumstances is a trivial consequence of the fact that prokaryotes are proliferating systems. The development of variety is not trivial, the less so, as it has been mentioned earlier that variety in principle may come about by two mechanisms: the *de novo* formation of the different systems and the continuous development of differences through a series of generations by non-identical replication. When describing the formation of prebiotic chemotons it has been emphasized that the metabolic network shown is only one of the possible variants, as the chemoton model is an abstract model that allows for the organization of every reaction system into a proliferating unit independently of its chemical nature, the only requirement being that should satisfy the conditions set by the abstract model. For example, the formose reaction is not the only autocatalytic cycle (replicative chemical cycle). The Calvin cycle or the reductive carboxylic acid cycle based in the transformation of organic acids are also similar cycles, and they not only may be, but actually are the bases for different types of metabolism in various living beings. No particular research work has been carried out concerning them so far, it is probable that a number of them will be discovered by a future systematic work, especially among archaebacteria. Hence the probability of the *de novo* formation of prebiotic proliferating systems with different types of metabolism is quite high.

The formation of different kinds through a series of non-identical replication does not need to be proved here, as this has already been done in the chapter dealing with prebiotic evolution. It may be open to debate how important the role of individual mechanisms has been in the development of the variety of prokaryotes, but it is indisputable that this variety has developed, and also that, in the case of prokaryotes, this phenomenon is essentially related to differences in the metabolic pathways, i.e., to differences that can be characterized quantitatively by stoichiometry. Thus condition 1 and the corresponding part of condition 7 are satisfied.

In the assembly of prokaryotes, between the individuals (i.e., the components of the assembly) stoichiometric interactions operate, even if the assembly consists of identical individuals. This is the consequence of the fact that the same finite nutrient pool is available for the individual cells, hence they compete for nutrients. The end products of their metabolism are also discharged into the same, finite pool, thus the increase in the concentration of metabolites exerts an inhibitory effect on the cells. Consequently, condition 2 and the corresponding part of condition 7 are fulfilled even for a homologous population of prokaryotes.

In the assembly of the variety of prokaryotes additional stoichiometric interactions are also possible. The metabolite of one prokaryote may be the raw material for another and vice versa. Thus even chains of interactions may develop between the various kinds of prokaryotes. For example, some prokaryotes decompose sugars into alcohols (*Clostridium* species, *Zymomonas*, *Thermoanaerobacter* species), others transform alcohols into acetates (acetic-acid bacteria), whereas numerous bacteria oxidize acetates into CO_2. This is only an example from the great variety of possible conversion chains.

The most important among these stoichiometric interaction chains are the cyclic ones, which ensure a continuous circulation of matter between prokaryotes, thereby preventing the total consumption of nutrients. Such chemical cycles and even several of their variations are known for practically all biogenic chemical elements. (This satisfies condition 3.) Best known are the reactions of the carbon and nitrogen cycles. The stoichiometric equations for individual reactions of such cycles are at present usually given only by classical stoichiometry, thus they do not contain the metabolism of prokaryotes performing the transformations (just like biochemistry omits the enzymes from the stoichiometric equations of enzymatic reactions). It is obvious that if the cyclic equations for the metabolism of individual prokaryotes will be elaborated, the interactions between the different prokaryotes will be describable quantitatively including both linear and cyclic interactions (fulfillment of condition 7). This also provides a possibility for determining in advance, for which qualitative and quantitative networks of prokaryotes is the realization of a continuous circulation of nutrient components stoichiometrically possible. In the long run, a stable prokaryote community can only exist within the frame of stoichiometric possibilities, but these possibilities are further restricted by kinetic and structural limitations. Our present knowledge of the living world shows that even so, a number of possible network variants exist.

Hence these networks develop in a "component system" the components of which are different kinds of prokaryotes, and the interactions between them are stoichiometric interlockings, i.e., the metabolite of one kind of the components serves as a nutrient for the other one. The interaction becomes a function, the function of feeding each other. As the component prokaryotes are proliferating systems, for these cyclic networks the condition concerning autocatalytic precursors is satisfied that the cycle (network) is of a component producing nature. The other requirement for AGSP is fulfilled in a trivial way from the thermodynamic viewpoint, because the cycle can operate only by consuming energy, otherwise it would be a perpetuum mobile. Thus these networks are of the same time autogenetic system precursors, which mean the satisfaction of condition 4.

Condition 5 is also fulfilled in a trivial way, because biological cycles for the transformation of matter can only exist if at their opposite sides opposite processes take place: if in one part organic material is synthesized, in an other part decomposition should take place; if in one part oxidation is performed, in another one reduction should occur, etc. However, it is not trivial but follows from the internal organization of the components that prokaryotes pass through an optimization process for performing these differentiated functions, and therefore they are capable of performing their task within the cycle or network with an increasing efficiency.

Thus it is clear that the assembly of prokaryotes is indeed suitable for being transformed into a 0-system in which processes can start that are needed for the production of components required by the next level of organizational processes. The experiences show that these organizational processes in the Earth have in fact taken place. There were already a number of cycles operating on the surface of Earth billions of years ago and they had to be brought into harmony to ensure the existence of a stable prokaryotic living world. The carbon and nitrogen cycles in themselves are not enough for the stable existence of prokaryotes; they should operate in accordance with, and coupled to, the other cycles (sulfur, oxygen, phosphorus, etc.). The supercycles of this organizational level had to be formed of necessity and also on a global level. In principle, these supercycles can also be described quantitatively in terms of cycle stoichiometry.

Evolution to eukaryotes

The circulation (operation) rate of cycles (networks) and supercycles (supersystems) organized from prokaryotes as components are determined by the following factors:

1. Intensity of the energy current operating the system
2. Efficiency of energy capture
3. Rate of chemical reactions taking place in the components of the cycle (prokaryotes)
4. Rate of material transport between the components

The intensity of the energy current operating the system is in the majority of cases independent of the system itself. The operation of the living world is based in light energy; the intensity of solar energy on the surfacc of thc Earth is a factor that may only be decreased and not increased by the living world.

Energy capture as well as chemical reactions are performed by the internal mechanism of the components (prokaryotes). In the course of evolution these processes became optimized, moreover, by the evolution of enzymes the individual reaction steps of metabolism became accelerated to such an extent that the transformation of nutrients into metabolic components took place immediately after they had entered the cell.

Hence for the acceleration of the circulation of prokaryote networks, factors 1–3 (intensity of energy current, efficiency of energy capture, and rate of metabolic reactions) cannot provide sufficient possibilities after the run-up of the evolution of prokaryotes. However, the increase in the rate of material transport offers an unutilized possibility. Material transport may occur by convection or diffusion. Both the extent and direction of convection change in space and time and are independent of the system, thus they cannot be influenced by the evolution of the system. On the other hand, the rate of diffusion is a function of the concentration gradient, which, in turn, is determined by the differences of the concentrations of the compound performing stoichiometric coupling and by the distance between the components. In productive prokaryotes the coupling compound is present in a maximum concentration, whereas it is practically absent in the consuming one, because here it transforms immediately after transport. Thus an increase in the concentration difference cannot be regarded as a factor considerably increasing the rate of diffusion.

Therefore it seems that for increasing the circulation rate of cycles and networks consisting of prokaryotes as components, the main factor may be the decreasing distance between the components. Because the proliferation rate of the components (prokaryotes) is determined by the rate of the cycle, the distance between the components has, in this system, a replicative function in terms of Csányi's theory, i.e., prokaryotes from which cycles are built in which the geometrical distance between the components can be decreased will have a proliferation advantage.

Thus from the chemical point of view, the geometrical distance between the "component neighbors" becomes an evolutional driving force. Therefore if some prokaryote, capable of displacement, can on purpose decrease this distance, e.g. by chemotaxis or indirectly e.g. by magnetotaxis (as magnetobacteria), this property of theirs becomes a replicative function in the system.

The distance between the two "component neighbors" is the smallest if the two prokaryotes are in direct contact. For this case, biology has a special term: symbiosis. As the validity of stoichiometric equations is not influenced by the diffusion distance of the diffusing compounds, the cyclic equations describing the stoichiometry of cycles (networks) organized from component prokaryotes are valid also for components being in symbiosis.

Direct contact is not the ultimate limit of closeness. The two prokaryotic cells can also fuse and this must have been a very frequent event in the phase of evolution in which the cell walls of prokaryotic cells had not yet been developed. In fusion, the two components fully coalesce, thus the composition of the resultant system is obtained by the summation of the terms in brackets on the left side of the overall cycle equations of individual prokaryotes. However, it is not at all certain that the stoichiometric description of the operation of the fused system is given by the equation obtained by adding up the overall equations of the two prokaryotes. Moreover, it is highly probable that a system corresponding to this equation is not even capable of functioning! The reason for this is that the metabolism of individual prokaryotes is a stoichiometrically closed system except for the intake of nutrients and secretion of metabolic products. However, when two non-identical metabolic networks unite, the intermediates of the two networks may also react with each other, thus the systems are no longer closed and it is not at all certain that between the reaction pathways developed in the fused system the stoichiokinetic conditions for a complete coupling described in detail in Chapter III are satisfied, which is the precondition for the stoichiometric stability of the system. The two, fused metabolic networks may contain identical, different (but not antagonistic), or antagonistic processes alike. It is necessary for the stability of the resultant prokaryote that the two fusing metabolic networks should form a single closed metabolic network. Presumably in the possession of the stoichiometric equations of the two prokaryotes it will be possible to decide theoretically in advance whether this is stoichiometrically possible at all and if so, which versions of the networks are stable. It has to be possible to calculate from the equations which processes have to be stoichiometrically eliminated, which new chemical transformations have to appear, and what stoichiokinetic changes have to take place in the old processes for the two fusing metabolic networks to transform into a stoichiometrically stable, uniform system.

At this point a prediction is possible. If the above reasoning is correct, part of the prokaryotes (perhaps even the larger part?) developed by the fusion of simpler prokaryotes having different metabolic networks, a fact that should be reflected in their metabolic network. Thus it may be supposed that if we become acquainted with the metabolic networks of different prokaryotes in sufficient detail, it will be possible not only to derive the genealogical trees by following the changes in the metabolic networks, but also to detect the polyphyletic origin of prokaryotes and the sites of fusions in these trees.

The fusion of prokaryotes does not mean that the systems thus formed are the elementary systems of the next level of organizational hierarchy, because in the resultant systems the independence, the "component" character of the prokaryotes ceases to exist. The new system is not a system of a higher organizational level formed by the collaboration of the two component prokaryotes (e.g., a symbiotic system), but rather a system of the same organizational level only having a more complex one, capable of more types of functions.

There is, however, a type of ultimate decrease in distance that leads in to the next level of organizational hierarchy, i.e., to the eukaryotic level, and this is the so-called endosymbiosis. It seems – as has occurred so often in the history of science – that biology put a question in the opposite way: it wanted to discover the origin of eukaryotes and not the evolution of prokaryotes, i.e., it proceeded again from the more complex phenomena to simpler ones. However, the extraordinary complexity and variety of the complex systems render their survey almost impossible, which is a well reflected fact in the immense jungle of publications dealing with the origin of eukaryotic cells. Nevertheless, from this vast amount of publications, three works can be mentioned that on the one hand may serve as very good points of orientation and, in the other hand, built onto each other, mark a uniform logical way for the interpretation of the course of events.

The first one is the far-reaching and provocative work of Margulis (1970), in which she laid down the Serial Endosymbiosis Theory (SET), supported by a number of sporadically published works declaring the endosymbiotic origin of chloroplasts and mitochondria, which again brought the question of the origin of eukaryotic cells into the focus of biological research. According to this theory, the organelles of eukaryotes (or at least part of them) were originally independent prokaryotes that entered into a host organism and lived in endosymbiosis with it thereby losing a part of their functions and thus becoming incapable of independent existence. According to Margulis, as a first step, an aerobic prokaryote of respirative metabolism was built into an anaerobic prokaryote having originally fermentative metabolism (from this process the mitochondria were formed), subsequently a cyanobacterium capable of photosynthesis was incorporated (this resulted in the formation of chloroplasts in plant cells). Similarly, the endosymbiotic origin of other cell organelles can also be rendered probable.

The second important orientating work is that of F. J. R. Taylor (1974), who noticed that the taxonomic confusion of unicellulars can be clarified on the basis of the types of genomes found in them. He calls cells with only one type of genomes *monads*. These are essentially prokaryotes. Cells having two different types of genomes, e.g., those having both a nucleus and mitochondria, are called *diads*, those with three types of genomes are called *triads* (nucleus, mitochondria, chloroplasts), etc. Multicellular plants are considered to be *polytriads* and multicellular animals to be *polydiads*.

The third outstanding work is that by Schwemmler (1979) who created a "periodic table" of cells. This system is in fact similar to the periodic table of chemical elements; it consists of periods and columns, the number of genomes increasing along the columns (though not completely consequently), whereas in the periods the variants of metabolism are repeated from fermentative to photosynthetic ones. Nearly two-thirds of the places in the table are occupied, the other one-third being unoccupied as yet. However, this periodic system also determines certain basic structural and metabolic properties for the unoccupied places; hence the system has very great predictive power as well. The table also resembles the periodic system of chemical elements in that it is constructed mostly empirically and probably a similar exact theoretical background will found for it as exists for the periodic table of chemical elements.

The latter two works, though tacitly, suggest: the primary question is not the sequence of evolutional events but rather what organizational modes may exist at all at this level of organizational hierarchy, and it is only of secondary importance whether they are in fact realized and if so, in what manner and what time sequence. It is an empirical fact that at this level of organization (at the level of eukaryotic cells) the organization of morphological structures also exists, but it is also indisputable that no morphological structures can be formed that are not made possible by the metabolism of the cells, the concerted system of the metabolic network of cell organelles.

Thus organizational possibilities are also determined at this level primarily by the harmony of metabolic networks while the organizational mode of the morphological structure sets further restrictions and variation possibilities only secondarily, within the limits of chemical laws.

Endosymbiosis is a transition between symbiosis and fusion from the viewpoint of stoichiometry: the metabolic networks do not coalesce to such an extent as in fusion, but neither do they keep their closed nature as in exosymbiosis. The metabolic network of the host cell has a direct connection with every metabolic product of the built-in cell, even with those that in the case of exosymbiosis would not penetrate its cellular membrane. If among them there are ones capable of reacting with the intermediates of the metabolic network of the host cell, the metabolic network has to be rearranged to reach a stable state.

We have to notice that endosymbiosis satisfies the last criterion of Csányi's autogenetic model: the interacting components (prokaryotes) are being compartmentalized. Thus self-reproducing (proliferating) systems develop that are built of prokaryotes as components. They may become the organizational elements, the components for the next level of organizational hierarchy. If a number of variants can be formed from them, the assembly of their varieties may again become a 0-system and organizational processes governed by evolution may start, i.e., the evolution of eukaryotes can take place.

Evolution of eukaryotes

Prokaryotes living in endosymbiosis are still not eukaryotes. Endosymbiosis, however, opens up many possibilities for organization. In the course of evolution the modes of organization have to pass through optimization processes with respect to both chemical and structural aspects in order to develop the most competitive versions under the given conditions. The fact that other compartments are situated within the compartment (the membrane of the host cell) indicates that the processes are also organized in space, thus the organization of morphological structures should necessarily play a decisive role. However, this is outside the frame of this work, and therefore will not be dealt with in detail despite the fact that the question of the origin of eukaryotes is at present attempted to be decided mainly by the study of morphological structures. For us, it is only important that these organizational processes have their own stoichiometric criteria and no system may exist at the eukaryotic level, if it does not satisfy these stoichiometric requirements.

Parts of the metabolic networks of two prokaryotes living in endosymbiosis are identical or similar. Obviously, in this case it is more favorable if these processes take place in the plasma of the host cell, because this is in contact with the nutrient through only one membrane and thus the processes can occur here with a higher rate than in the built-in cell which is separated from the nutrients by its own membrane, and also by the cytoplasm and membrane of the host cell. For this, the transfer of the appropriate genes, or at least the multiplication of the corresponding genes of the host cell, or the elimination of these genes in the built-in cell is necessary. In this way the genome of the host cell increases, whereas that of the built-in cell decreases. Presumably, this increase in the host genome initiated the development of the morphological structure of the genome, i.e., the formation of chromosomes and the nucleus.

In symbionts, processes may also take place that produce the same product via different mechanisms and with different efficiency. An outstanding example of this is the regenerative synthesis of ATP, which produces 2 ATP molecules by the fermentative metabolic pathway and 28 ATP molecules through the oxidative mechanism by the consumption of 1 molecule of glucose. It is obviously more favorable in the case of an endosymbiotic pair consisting of a fermentative and an oxidative member, if the oxidative partner is specialized to the regeneration of ATP.

In the course of such rearrangements of metabolic processes the dependence of the partners on each other increases to such an extent that the partners may completely lose their independence, i.e., the built-in cell becomes an organelle, and the host cell transforms into a complex eukaryote. In principle, it seems possible to determine theoretically the stoichiometrically stable network variants in an exact way by means of an appropriately advanced cycle stoichiometry in the detailed knowledge of the metabolisms of the given symbionts, i.e., to decide what processes may be transferred, which ones have to be eventually eliminated, and how the resultant network has to be organized stoichiometrically in order to ensure a stoichiometrically stable network of the resulting eukaryotes.

The periodic system of Schwemmler contains 80 positions. These positions have been obtained essentially by the variation of the basic types of metabolism and the number of genomes (by taking into account also some other aspects). This in itself proves the large number of possible eukaryote types. In fact, not only the basic types of metabolism (fermentation, respiration, photosynthesis, anaerobic, aerobic), but also the variety of reaction pathways within these processes contribute to variations, hence by orders of magnitude more variations may exist than this number also on a mere stoichiometric basis. However, the variety of eukaryotes is increased not only by stoichiometric factors, but also by morphological and structural factors. Thus, as is proved by experience, an extraordinary variety of eukaryotes may be, and is in fact, formed in the course of evolution. Hence the eukaryotes can also form a 0-system in which they are the components. Interactions may exist between them, not only stoichiometric ones (of course, these too), but distinctly biological ones as well (e.g., a predator –prey relationship).

Thus the process described by the autogenetic model can also be repeated at the level of eukaryotes and as a result, systems organized from eukaryotes as components appear; they are Taylor's polydiads and polytriads, i.e., multicellular animals and plants. In the organization of these systems the main roles are played by new, biological, regulating, and information-carrying interactions together with morphological and structural possibilities and limitations. However, in the background stoichiometric correlations are also present: the somatic cells from which multicellular beings are constructed are also in metabolic relation with each other, and a division of labor of a stoichiometric nature is found between them. Cycle stoichiometry is also indispensable in the quantitative description of these correlations.

Finally, the variety of multicellulars may also form a 0-system. There are interactions operating between them, and they also can be organized into systems, into supraindividual systems. These systems can be divided into two large groups: systems of ethological type and those of ecological nature. In the former group there is no stoichiometric relationship between the components (i.e., individual), or at least, this is not characteristic of them, the system is held together basically by the informational relationship. On the other hand, systems of the ecological types are also based on stoichiometry and thus stoichiometric equations can be applied in their description.

From all of this it is apparent that stoichiometric organization runs through all levels of the organizational hierarchy of the living world. Of course, by the progression of evolution several further interactions appear and become predominant. Chemoton theory does not intend to discuss these interactions, but in their background there is always stoichiometric organization which is a limiting factor for their operation, which is at the same time also the mechanism of function for other interactions (structural, hormonal, neural, informational, etc.) and which may be described by cycle stoichiometry (including stoichiokinetics) developed in the frame of the chemoton theory. It was possible to write in this sense at the beginning of this volume that our aim is to create the *basis* for an exact, quantitative biology: to produce a basic model – the chemoton model – from which, and a methodology – cycle stoichiometry – by which, the phenomena of the living world can be derived both qualitatively and quantitatively.

Applications in Biotechnology

Limits of the prospective development of biotechnology

"Biotechnological processes" are probably as old as human societies are, and even industrial biotechnologies have existed at least for a century. However, there is a basic difference between these and what we may call "new biotechnology". Presumably at least such basic differences will exist between the "new biotechnologies" of our days and those after the turn of the twentieth.

"Classical" biotechnological processes were always based on empirical findings and used the whole living system, be this living system bacteria, molds, yeast cells, or eventually a mammal, as e.g., in the production of vaccines. "New" biotechnological processes are usually targeted at a specific objective based on the results of biochemistry or molecular biology as well as in abstract, theoretical knowledge. New biotechnology operates with submicroscopic "machines" and mechanisms: genes, gene transfer, specific macromolecular mechanisms, enzymes, enzymatic reactions rather than with complete living systems, as opposed to the old ones.

An individual living system contains several thousand genes and specific macromolecules, types of enzymes and at least the same number of enzymatic reaction steps, and macromolecular mechanisms that operate in harmony, in a controlled, even program-controlled, way. Thus the above-mentioned submicroscopic mechanisms can be organized into regulated, program-controlled productive systems. The characteristic of biotechnological processes after the turn of the millennium will be the designed realization of such systems and their introduction into production. The difference between the "new" biotechnology and that after the turn of the millennium may be compared with the difference between production by machines and industrial production using automated robot technology.

"New" biotechnology was started by our ability to find methods for the aimed selection of individual genes from among thousands of genes present in the cell, to determine their sequences, to multiply these individual genes in the laboratory (cloning), and to build these genes into the genetic material of bacteria thereby forcing them to produce proteins determined by this implanted gene, independently of whether the bacteria themselves may use this protein or not.

Hence, whereas "classical" biotechnology has selected and utilized those having a productive program utilizable for human society from among the program-controlled systems of the living world the "new" biotechnological processes can even modify the program of these systems. This naturally promises unprecedented possibilities, but at the same time, also clearly shows the limitations restricting these possibilities for biotechnological manipulations as a result of our insufficient knowledge.

One of the limiting factors is that although we can unambiguously predict the direct consequence of successfully introducing the given gene (the protein corresponding to this particular gene will be produced in the cell), the indirect consequences cannot be foreseen and predicted, even approximately; what changes will occur in the metabolism of the cell, how the importance of individual metabolic pathways will be shifted, what the appearance of the new protein or enzyme means for the life, properties, stability, and vitality of the cell.

The situation is similar in the case of the so-called hybrid cells originating from cell fusion. In these, both genetic materials and cell plasmas are present after the fusion. However, the properties are not additive: the hybrid cell is only viable if a uniformly operating, regulated system that is program-controlled by the genetic material can form from the mixture. For this, part of the genetic material should be eliminated, i.e., unnecessary or even harmful commands should disappear, together with all of the operations and processes that do not fit into the cell plasma of the new system. At present we cannot tell which genes will disappear and which will not, what the fused system will be like, or even whether the hybrid cell resulting from the fusion is viable and capable of proliferation or not.

Another limiting factor in the development of biotechnology is the molecular genetic approach the restricting effect of which in the cognition of the origin and nature of life was discussed earlier. As discussed in detail in the well-known paradigm theory of Kuhn, recognitions enabling revolutionary changes to occur in science have to fight against the old theories, but after their wide acceptance they begin to obstruct the development of newer discoveries of even greater impact. In the second half of the 1950s and during the 1960s molecular biology made a breakthrough by placing nucleic acids and their sequences in the center of biological research and thinking. As a consequence, the majority of biochemical and genetic laboratories having some weight turned their attention toward nucleic acid-related research. Today the biologists think in term of genes, nucleic acids, and protein sequences, sometimes even identifying life with genes, and forgetting that genes are mere technological commands and that enzymes made by genetic commands are in the last analysis only dedicated machines of molecular dimension for carrying out elemental operations on individual types of molecules.

As has been shown throughout this book, living systems are not identical either with technological commands or with the assembly of the variety of specialized molecular machines. Enzymes, as such machines, can be organized into technological "assembly lines." The cell is a system composed of a variety of such assembly lines controlled in space and time by programs contained in the genes. Today's biology cannot operate in terms of assembly systems that it cannot treat and describe quantitatively. The disadvantages of this are already apparent: the promising long-awaited "prodigy" of gene technology, the "self-nitrogen-fertilizing" plants that would be obtained by the transplantation of nitrogen binding genes has not yet been born on even a laboratory scale, in spite of concentrated research for decades. Investigators come up against an invisible wall and start to guess that the transfer of genes, i.e., the technological commands, is by itself insufficient. Scientific workers dealing with cell fusion meet with similar difficulties.

Still, the really great promises of biotechnology are such complex productive systems: a chemical industry manufacturing by series of enzymatic reactions without harmful wastes, electric energy and hydrogen fuel produced by photosynthetic systems, the correction or compensation of broken-down metabolic systems in health control. However, in recent years, all of the significant research and development institutes of the world have chosen molecular biology-related subjects in order to enjoy the financial benefits of the biotechnological revolution. Therefore research capacity will be engaged for the next 5–15 years in the field of macromolecules and the corresponding genetic manipulations. This takes the intellectual and financial means away from the research on molecular process systems, and thus becomes an obstacle in the way of development.

One-step chemical syntheses

To carry out chemical syntheses and chemical transformations, industry today is forced to use very reactive, aggressive reagents. These reagents, as well as the by-products of the reactions are harmful to the environment and poisonous to man. The most serious and almost insoluble problems of chemical industry originate from this fact: the disposal of dangerous wastes, environmental and even human tragedies and mass catastrophes. This problem exists at the industrial scale even in the synthesis of compounds that are readily synthesized in living systems by means of enzymes without using and producing harmful materials. Thus the expectation of carrying out such reactions in enzymatic syntheses also in the chemical industry is justified, the more so as enzyme reactions are usually substrate- and reaction-specific, i.e., by their application side reactions can be eliminated and undesired by-product formation can be avoided. Moreover, in the production of chiral compounds, in enzymatic reactions usually only the desired optical isomer is formed, this being an almost unattainable dream of organic chemists.

In spite of this, the organic chemical and pharmaceutical industries hardly ever use enzymes in syntheses, the only exception being the food industry, which applies several enzymes in its production (rennin, glycose-fructose isomerase, catalase, amylase, etc.). The reason for this is mainly that enzymes can only be produced at present from living organisms in a very complicated way, with a poor yield and high cost. This obstacle seems to be surmountable: on the one hand, in current technology the preparation of cheap enzymes is not unprecedented, even in great amounts (washing powder enzymes), and on the other hand, gene technology enables us to make bacteria produce almost arbitrary enzymes with a high yield. What is not yet solved is the digestion of the bacterial cells for the liberation of the enzymes from the cell at an industrial scale, but this is also to be expected if larger intellectual and financial investment is made in this field.

The other reason why enzymes have not found wide industrial application is that although enzymes can theoretically be used an infinite number of times for the acceleration of reactions, in practice, they cannot be recovered from the reaction mixture and thus they are wasted after a single use. This can be helped by binding the enzymes to solid supports or by their inclusion onto gels. These solutions have already been introduced into practice in more than one way, even at an industrial scale (e.g., aminoacylase, glycamylase, glycoseisomerase, hydantoinase, lactase, nitrilase, penicillin G acylase, penicillin V acylase).

All of this is essentially the field of "classical" biochemistry. Their combination partly with gene technology and partly with organic chemistry, however, promises revolutionary changes. First, by the application of gene technology, not only can the yield of enzyme production be multiplied but also the nature of the enzyme can be modified: the sequence of the implanted gene can be changed and thus also the sequence and thereby the spatial structure of the protein coded by the gene. This results in a change in the reactivity and resistance of the protein. For example, the Cetus Corporation has modified the three enzymes used in the production of glycols so that the pH optimum of the three enzymes, originally operating in a pH interval of 3, became identical, thus they function in the same medium simultaneously. The Genex Company uses a similar procedure for the bacterial production of phenylalanine (Maugh, 1984). By organic chemical modifications, the active sites of enzymes can be changed thereby imparting completely new catalytic properties to the enzyme (Kaiser and Lawrence, 1984).

Several years ago the thought that not only proteins could function as enzymes was quite incredible; however, the prediction, and later on the discovery, of enzyme RNAsmade this thinking untenable as was discussed earlier. Organic chemistry has also come closer to the preparation of synthetic enzymes; cryptates (Lehn, 1985) and cavitands (Breslow et al., 1983; Cram, 1983) are significant steps toward the preparation of synthetic enzymes. Now it is obvious that enzyme functions are not the privilege of proteins; in principle, enzymes can be prepared from each macromolecule, having a sequence-dependent tertiary structure. Because of its complexity, the design of such a synthetic enzyme was unthinkable earlier. It became possible by the advent of computerized three-dimensional design of molecular structures such as globular macromolecules. At present, this is mainly a question of computer capacity. Owing to the extraordinary advantages and the pressure of environment protection on the chemical industry, it is to be expected that enzymes, even synthetic ones, will be applied in the chemical industry in the last decade of this century, which will be followed by a radical transformation in the structure and technology of the chemical industry.

However, as has been shown in detail, stoichiometry cannot calculate with enzymes. On the other hand, cycle stoichiometry enables us to calculate enzymatic reactions algebraically. This, however, is not yet of any practical importance in the case of one-step enzyme reactions.

Chemical syntheses by reaction networks

One of the most important problems in organic chemical syntheses is that in the case of multi-step synthesis pathways, intermediates have to be isolated from the reaction mixture in a more or less pure state, usually after each reaction step, and the next step has to be carried out in a new reaction medium. This is not only very energy- and time-consuming, but the losses in yields are also accumulated step by step. In living systems, even very complex reaction series occur continuously, in the same medium without the isolation of individual intermediates, thus the yield is very high, practically 100%. It is quite obvious how important it would be for the chemical industry to have technologies in which the initial materials are introduced into a reactor (e.g., a column filled with a synthetic resin support) at ambient temperature, and at the other end of the reactor the solution of the ready product leaves, the complex synthesis consisting of 10–20 reaction steps having meanwhile taken place. Such systems can be realized even at present by means of enzymes, moreover, by binding synthetic enzymes to supports an extraordinary variety of chemical products will be produced in the future, including those not occurring in the living world.

As realized examples, first those procedures should be mentioned that were already discussed in detail in Chapter III. All of these processes utilize the glycolytic enzyme system of yeast and regulate the operation of this enzyme system by substrate addition, cofactor regeneration, enzyme inhibition, etc., so that it would produce the desired end product. This is how phosphoenolic pyroracemic acid (Gánti and Csóka, 1975), fructose-1,6-biphosphate (Neuberg and Lustig, 1942; Gánti, 1975), 3-phosphoglycerol acid (Mandl and Neuberg, 1952), adenosine-5'-monophosphate (Ostern and Terszakowec, 1937), adenosine-5'-biphosphate (Gánti et al., 1970), and adenosine-5'-triphosphate (Ostern Baranowsky, and Terszakowec, 1938; Ostern, Terszakowec, and St. Hubl, 1938) could all be produced. In these procedures 2– 20 enzymes act simultaneously, i.e., the same number of reaction steps take place at the same time, at ambient temperatures. The often very complex reaction network produces the end product without the need for the isolation of intermediates.

However, these syntheses were designed in a semi-empirical way: the qualitative design of reaction networks was possible on the basis of existing biochemical knowledge, but no quantitative projection was feasible. Enzymes are present in these systems in dissolved state, usually as simple plasmolysates. Nevertheless, it would be possible to carry out all of this also by pure, bounded enzymes, in which case a quantitative design would be necessary, not only for coordinating enzyme capacities with each other, but also because in these systems (except for PEP production) coenzyme regeneration cycles are also present, thus the activation of the enzymes should be brought into harmony not only with substrate concentrations and intermediates, but also with the regeneration of the coenzymes (ATP, NAD). This is made possible by cycle stoichiometry: the amounts of the components in the system should be chosen so that their stoichiokinetic coefficients are identical. By knowing the turnover numbers of the supported enzymes – which may be indicators for a warranted quality in industrial products – this can be performed by a simple calculation without the need for solving complicated, nonlinear differential equation

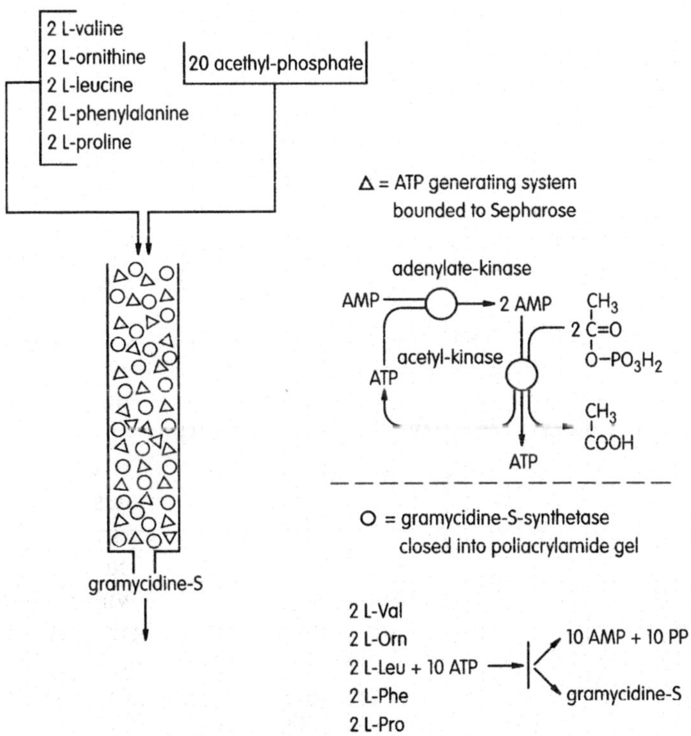

Fig. 11.4. The synthesis of gramicidin S by Wang et al.

systems for the kinetic evaluation.

Development toward the industrial utilization of reaction systems applying bounded enzymes has already started: e.g., Wang and co-workers accomplished the synthesis of gramicidin S in such a way. They obtained the ring decapeptide, the antibiotic gramicidin S, with a yield of 80% from amino acids introduced into an enzyme reactor containing gramicidin S synthetase included in a polyacrylamide gel, moreover adenylate kinase and acetyl kinase performing the regeneration of ATP bound to Sepharose. The synthesis is illustrated in Fig. 11.4 (cited by Szentirmai, 1986).

Synthesis of Living Systems

The objective of synthesis

In his famous book, "The Biological Time-bomb," G. R. Taylor (1968) writes that Professor Charles Price, as the newly elected president of the American Chemical Society, in 1965 publicly proposed that the synthesis of life should be made the national objective of America. He suggested a 20-year program that – according to his plan – would employ a quarter to a half of the experts in the United States. If this program had started then, it would have ended by now. It is of interest to investigate on the basis of our present knowledge, how reasonable this program would have been, and whether it would have ended successfully.

Professor Price thought that the work should start with the synthesis of the important constituents of living organisms. In the next step, viruses would have been synthesized, then the subunits of which cells are built, after that, complete cells, and finally multicellular beings. "I feel," Price told Taylor, "that in this country chemists need the stimulating force of a great objective to which they can contribute with their work. They need a goal which encourages them and fills them with the sensation of a specific task, similarly as the landing of a man on the Moon a stimulated technical and electronics research." Of course, as the program has never been started, its success cannot be judged. As will be shown later, although it cannot be decided at present whether or not living systems have been born during laboratory experiments, it can definitely be stated that if the program of Price had been started, artificial production of living systems would have been accomplished, even if not at the multicellular or eukaryotic level. What can justify such a definite statement?

First, it should be established that almost any component of living systems can be prepared synthetically at present: genes, ribonucleic acids, enzymes, hormones, let alone the small molecular compounds present in living organisms. Membranes of biological character, enzyme complexes can also be synthesized, as well as virus nucleic acids, virus proteins, or even complete viruses. What cannot be produced synthetically as yet are the skeletal materials produced by the spatial arrangement of macromolecules: cellular membranes, cellulose fibers, chitin, etc. Thus today we would not be able to produce all of the components of any living being synthetically, but there is hardly any theoretical obstacle to do that with today's simplest living organisms (e.g., microplasms, thermoplasms). Truly, that task will engage a very large research capacity, but within a few years, it would practically represent routine research.

The question is whether this is what the artificial production of living system means. To the public this means in general much more, to science much less. Public opinion expects the chemical synthesis of a homunculus, or at least some moving, running being in a test tube. Even Taylor writes about the program of Price that "... in the next step we would synthesize, viruses, then subunits of which cells are built, then whole cells and finally multicellular organisms."

However, it must be clear that multicellular organisms cannot be synthesized even theoretically from chemical compounds; at most they can be constructed from synthesized cells. Moreover, this cannot be done either, because for this, it would be necessary to synthesize an infinitely large number and kind of cells separately and somehow bring them into synchronized operation. Even this is impracticable. However, it is totally impossible to create the microscopic structures by assembling cells the billions of which are required to form e.g., a mammal. Only one way is feasible: to let the multicellular organism be formed by the usual way of ontogenesis from the appropriate cell carrying the complete program of multicellulars.

Naturally, this cell proper could be synthesized. In this case, the multicellular developed from it should also be considered to have been made synthetically. At present, the theoretical possibility of the production of such a cell cannot be rejected outright. The question is rather why the artificial synthesis of such a cell would be necessary. What can be the reason for the artificial synthesis of life at all? Taylor writes:

> ... as soon as it becomes possible, there will be biologists trying `to produce life', not only for the sake of bother, but also because the most reliable test of methods and theories is if they also work in practice. Whoever wants to understand the structure of a clock, can best check whether he has understood the principles, if the assembles a clock and sees if it works. In this, national matters of prestige may also play a role: it will certainly excite worldwide interest if

someone first produces living material. For this purpose, the production of a single cell capable of division and proliferation, taking nourishment, carrying out metabolism, secreting refuses and showing sensitivity towards its environment would be enough. It would be an astounding achievement!

In fact, just as in chemistry the only aim of the so-called total synthesis is the verification of a theoretical molecular model, in biology, the objective for the synthesis of living systems cannot be other than the justification of theories concerning the nature of life and the organization of living systems. It would be a very naive thing to believe that we can produce better living beings by an artificial synthesis than those populating the Earth today which has been improved in the 4 billion years of evolution, or that we can produce such perfect ones in a simpler, cheaper way than the biological one. Of course, we use even today living beings that are more suitable for performing particular tasks than those optimized by nature, but they are much simpler to produce by the modification of already existing living beings than by total synthesis. For this modification, gene technology provides us with suitable methods.

Neobiogenesis

This expression originates from Keosian (1968) and means the *de novo* genesis of life taking place at present. The assumption of neobiogenesis does not mean the denial of the experimental results of Pasteur, for contrary to public opinion, he proved by his experiments that all of the theories and experiments intending to prove neobiogenesis provided positive results only because of experimental errors, rather than that no living being can originate from inanimate matter.

From time to time publications appear in the scientific literature calling attention to phenomena regarded as neobiogenesis. A good survey of earlier such works is found in the book by Oparin (1957), whereas later observations are summarized in the book by Bahadur and Ranganayaki (1981). It has to be established that there is no single experiment or observation that could prove scientifically that the structures (mostly microscopic ones) formed are really living beings. But it has also to be noted that the opposite has not been proved either. It is by all means noteworthy how an extraordinarily strong tendency to develop structures is shown in solutions containing organic compounds. This is experienced not only by those dealing with neobiogenesis, but all those working with sterile solutions or media. The only difference is that the latter group only states that a "deposit" is formed and as this is their daily experience, they fail to ascribe any importance to it. In fact, the formation mechanism, composition, and properties of these fine deposits formed in various sterile solutions have not been studied as yet, except for some fanatics who think they have discovered neobiogenesis in these deposits. It cannot be decided at present whether they are right of wrong, because of the lack of suitable studies.

It is also true in this case that adequate progress is hindered mostly by the current attitude of researchers. How true this is may be well illustrated with the example of the real or presumed discovery of bacteria living at 250 °C, mentioned in Chapter IX. Trent and co-workers, whom Baross and Deming did not provide with the bacteria originating from deep sea springs and according to them, proliferating at 250 °C, repeated the experiments of Baross and co-workers under "sterile" conditions, i.e., they did not inoculate the medium used by Baross et al. by the bacteria in question. When they obtained results similar to those of Baross and Deming on the "inoculated" medium, their only conclusion was that in the experiments of Baross no heat-resistant bacteria were produced but some artifacts instead. However, it is presumably much more important to notice that in their experiment bacteria-like spheres of 1μm diameter were formed and that protein reagents gave positive results even to such an extent that the "protein concentration" increased from zero to 10^2–10^3 μg/ml in 2 –3 hours. So it seems that the process toward organization into a living system was started very quickly, though as a carbon source, only $NaHCO_3$ was present in the medium (Trent et al., 1984). Neither the structures formed, nor the composition and properties of the substances giving positive protein reaction have been studied more thoroughly.

From the point of view of neobiogenesis the already mentioned experiments of Fraser and Folsome are even more interesting. The authors studied the formation of the microscopic structures under conditions essentially similar to those of Miller in the original experiments: they subjected the mixture of N_2, CH_4, CO_2, and H_2O to the effect of UV irradiation. Their most important conclusions are: (1) CO_2 dissolved in water is incorporated into the organic material of microspheres by photosynthesis; (2) as an intermediate of photosynthesis, formaldehyde is present in the system; (3) the formation kinetics of microspheres is of an exponential nature (Fraser and Folsome, 1975; Folsome, 1976, 1977; Folsome and Brittain, 1981). Unfortunately, many questions of crucial importance for neobiogenesis are not answered in their works, such as: (1) Is the photosynthetic incorporation of carbon into the organic compounds bound to the microstructures formed or is it independent of those? (2) Does the formaldehyde formed play a decisive role in the development of microstructures? (3) Is the exponential growth in the amount of the microstructures a consequence of the operation of an autocatalytic reaction system one of the simple end products of which are the microstructures, or is the microstructure an autocatalyst in the reaction system, or is it a structure corresponding to the self-reproducing microspheres discussed earlier, in which case the exponential kinetics is the consequence of the successive divisions of the spheres subsequent to their growth? (4) What are the properties of these microspheres? Do they show any kind of metabolism? Do they grow? Do they divide? Etc.

Finally, we should get acquainted in more detail with the most interesting experimental research concerning neobiogenesis carried out by K. Bahadur and his wife, S. Ranganayaki, and their students and co-workers at the University of Allahabad in India for almost three decades. They noticed that if a sterile solution of formaldehyde, ammonium salts, and mineralic salts of biological importance containing molybdenum and iron as inorganic catalysts is exposed to the effect of sunlight, a mass consisting of microscopic, spherical particles of organic composition is formed in the solution. They called these particles Jeewanus, which is the Sanskrit word for "particles of life" (Bahadur et al., 1963, 1966; Bahadur and Ranganayaki, 1970, 1983; Bahadur, 1964, 1966, 1967). They have prepared and studied a number of varieties of Jeewanus over the past three decades.

Bahadur and his co-workers *believe* that Jeewanus live, i.e., that they are the simplest living systems. However, the scientific world *believes* that they are inanimate artifacts and it does not even consider the results, let alone tries to disprove them. Nevertheless, there is a significant difference between the two beliefs. The basis for the belief of Bahadur and co-workers is the experience and experimental observations of three decades. As a contrast, the belief of the scientific world is based on prejudice. It is prejudice, first against the unusual, unexpected, and strange experimental results, and second, against the modest, too simple, and hardly equipped experimental methods.

What are the experimental findings on which the belief of Bahadur and co-workers is based? First, the formation of microspheres: a large number of spheres with dimensions in the range of 0.5–1.0 μm are formed, part of them single, part of them double, another part seems to be budding, and finally there are some that are aggregated into small groups. Then, the complexity of the composition of the microspheres: amino acids, peptides, sugars, nucleic acid bases, aromatics, and phospholipids have been detected in them (Bahadur, 1954; Briggs, 1965; Raina, 1973; Singh, 1975; Kumar, 1976). The major part of amino acids, sugars, and nucleic acid bases is incorporated into macromolecules; they can be detected only after hydrolysis. According to Bahadur, the microspheres proliferate by budding so that buds appear at the surface of spheres, start to grow, and after reaching a certain size they become detached and leave the sphere. The microspheres are also internally structuralized. Catalase, peroxidase,

ATPase, phosphatase, and urease activities have also been found in them. The membrane surrounding the particles is of a phospholipid character and in the methanol or chloroform extracts of the particles, in addition to numerous other substrates, unidentified components, α-lecithin, and choline have been identified (Singh, 1975).

All of this already suggests that Jeewanus are not too distantly related to simplest living systems. This is so, even if doubt may arise concerning the reliability of some of the analytical methods used. It is essentially unimportant whether the component detected is in fact α-lecithin or 2-D-deoxyribose; what is significant is that all kinds and types of biological compounds produce a multitude of components in the microsphere. This, in turn, can be proved by experiments.

Bahadur's most decisive argument for the living character of Jeewanus seems to be that they metabolize. However, as unambiguous is the concept of metabolism for living beings at the present level of evolution, as ambiguous it becomes at the borderline of life. It is especially undefined as to what should be regarded as metabolism of biological nature in the different phases of the genesis of life. Oparin also spoke about some kind of metabolism for coacervates, although in that case probably only simple physical mass transport took place. S. Fox thinks also that the proteinoid microspheres "metabolize," and he is certainly right if we consider metabolism as the process in which the proteinoids that are present in the medium become incorporated into the membrane of the microspheres, thereby initiating the growth and budding of the microspheres. The metabolism of Jeewanus is more advanced than that, for they create their own material by chemical transformation, and although this transformation is not clarified so far in every detail, it has to occur via a necessarily complex reaction pathway, the raw materials being HCHO or CO_2, whereas the product is a multitude of complex organic compounds and macromolecules. If by this we mean metabolism, Bahadur's statement should be considered proven. If we mean more than that, Bahadur's statement may be correct but it cannot be regarded as proven.

The three main processes of the metabolism of Jeewanus – participation in the photosynthetic splitting of water, photosynthetic assimilation of CO_2, and photosynthetic fixation of nitrogen – are fairly well verified (Rao et al., 1980; Smith, Folsome, and Bahadur, 1981; Bahadur and Ranganayaki, 1983). In the Jeewanu, a ferredoxinlike substance is present that is capable of substituting ferredoxin in a chloroplast–ferredoxin–hydrogenase system in the splitting of water into H_2 and O_2. Interestingly, the Jeewanu has a definite nitrogenaselike activity. This is the more interesting, as the biological fixation of nitrogen is realized by a rather complex biochemical system, in the center of which there are two fairly large protein complexes containing molybdenum and iron atoms. This protein complex is capable of reducing the very stable nitrogen molecule to ammonia. Under certain conditions, nitrogenase reduces acetylene to ethylene; therefore, this property is utilized for measuring the activity of the system. A. Smith, C. Folsome, and K. Bahadur have shown that the Jeewanu suspension has nitrogenase activity and reduces acetylene to ethylene when subjected to both visible and UV irradiation. The photoreductive incorporation of atmospheric nitrogen into the organic structures has also been proved by a direct method (Smith et al., 1981). The same authors also proved that these microstructures can reduce CO_2 in a photosynthetic way and that the $^{14}CO_2$ present in the medium can be incorporated into the organic structures upon irradiation (Smith et al., 1981).

Possibility of an exact proof

If we were to prepare the simplest living cells of our days synthetically, e.g., a microplasm or a thermoplasm, we should proceed in the following way. First, we should clarify the complex metabolic map of the given cell. With its detailed knowledge, we should decide which compounds have to be placed into the system and which would automatically form during the operation of the system in the presence of the former ones. Subsequently, the nucleotide sequence of the complete genome should be determined, and after that, the sequences of the RNAs and proteins should be defined. The following step would be the determination of the composition and structure of the surrounding membrane, and then the preparation of the necessary small molecules would follow with the subsequent synthesis of DNAs, RNAs, and proteins of appropriate sequences. Finally, the solution containing all of these compounds should be enclosed in a membrane microsphere of suitable composition, structure and size.

Today we have no reason to assume that this system would not operate or live. There is no need either to breathe a "soul" into it, as the thermodynamic driving force, the difference in the free enthalpy content of initial materials and products, would automatically start the operation of the system. Whether this system is a living one or is not, can be best decided by comparing it with the sample according to which it has been synthesized. If they are identical in size, shape, structure, composition, and behavior, we should declare it a living system, for the sample of which this is a complete copy is considered to be a living system by common consent.

In principle, such a synthesis can now be performed. It is, however, another question whether it could be realized in practice because of the very high costs and large research capacity needed. Nevertheless, a national program of such volume as was suggested by Price in 1965 would have been able to solve this task.

However, this objective can be achieved in a much simpler way, if we follow the pathway shown by nature. All of the studies and all of the experiments performed so far suggest that matter is organized into living systems spontaneously and necessarily under suitable conditions. Thus it seems that if someone wants to synthesize living systems, all he has to do is to provide the suitable conditions and to let matter be organized into a living system. This is essentially the way that was followed by C. Folsome and K. Bahadur. And this was what J. B. S. Haldane recognized half a century ago when he gave the following recipe for the synthesis of living systems: "Take a planet with some carbon and oxygen, irradiate it with sunlight and cosmic radiation and leave it alone for several hundreds of million years" (Haldane, 1954). The diligent research work carried out in the past decades rightly fills us with the hope that there is no need for cosmic dimensions and geological periods of time to carry out this experiment.

However easy it seems to be to perform this task in that way, it is difficult to prove the result. In this way, living systems similar to those with which Earth is now populated would not develop; at most such ones that had been formed 4 billion years ago in Earth would develop. Thus, a direct comparison for proving the result is out of the question. There is no other way of doing this than to abstract the principal properties of different kinds of terrestrial living beings, which may serve to judge whether they are living or not. The setting of a definite borderline between living and inanimate systems cannot be avoided either, for in the opposite case it always remains a question of *belief* and not that of *certainty* whether the system formed lives or not. In addition, it is not enough to formulate this criterion verbally. It is not enough partly because the sense of the word is always disputable, and partly because the verbal formulation in itself does not provide the possibility for a quantitative checkup. Thus a model is necessary, an abstract model of simplest living beings, which would make it possible to decide by quantitative measurements whether the given system is living or not.

Hopefully, after having read this book, there is no doubt in the reader's mind that the chemoton model suits this purpose. It is not the only possible model; although the author does not know of any terrestrial living being that in its organizational basis contradicts the chemoton model. If, in turn, we accept the chemoton model as the minimum model of viable systems, we know exactly what we have to explore in a given system in order to decide whether it is living or not. On the basis of what is written in these two volumes, we can decide unambiguously by suitable quantitative measurements about e.g., the microspheres of Folsome or about the Jeewanus: (1) Are the microstructures inert end products? (2) Are they catalysts? (3) Are they autocatalysts? (4) Are they proliferating microspheres? Or (5) microspheres containing a hereditary program and operating program-controlled, i.e., are they systems of chemoton construction – are they living systems or not?

Variants 1–3 can easily be decided upon by simple measurements. To prove point 4, a serious effort of a well-equipped and financially strong laboratory would be necessary. To prove point 5, a modest national research program would be needed, the efforts of a single laboratory would not be sufficient, be this at Allahabad University or Harvard University. This is why we cannot decide at present whether living systems have already been synthesized or not. However, the chemoton model implies not only the possibility to decide whether a spontaneously formed system is living or not. As has already been shown in the design of a prebiotic chemoton, in the possession of the necessary chemical information it also provides means for the exact, quantitative design of living systems and enables us to carry out aimed experiments for the synthesis of living systems. The realization of such a program would be a real total synthesis of living systems.

References

AGAR, W. E. (1951): *A Contribution to the Living Organism.* Cambridge University Press, London.

AKABOSHI, M. and K. KAWA1 (1971): A biological synthesis of adenosine di- and triphosphates using recoiled ^{31}P atoms obtained from the β-decay of ^{31}Si. *Biochim. Biophys. Acta* 246, 194-200.

ALFVÉN, H. (1980a): *Cosmic Plasma.* Reidel, Dordrecht.

ALFVÉN, H. (1980b): Origin of the solar system. (in Hungarian). Lecture on the COSPAR Congress, Budapest, 1980. In: I. Abonyi (ed.), *Fizika* 1981–82. Gondolat, Budapest.

ALFVÉN, H. and G. ARRHENIUS (1976): Evolution of the solar system. NASA. SP 345, Washington.

ALFVÉN, H. and C. G. FÄLTHAMMAR (1963): *Cosmical Electrodynamics.* Oxford University Press, London.

ALLEN, W. V. and C. PONNAMPERUMA (1967): *Curr. Mod. Biol.* 1, 24.

ALMÁR, I. and T. GÁNTI (1981): Earth-independent colonies within the solar system. *Adv. Earth Oriented Appl. Space Technol,* I, 119-122.

ARVIDSON, R. E.; E. A. GUINESS; H. l. MOORE; J. TILLMAN and S. D. WALL (1983): Three Mars years: Viking Lander 1 imaging observations. *Science* 222, 463-468.

ASTBURY, W. T. and F. O. BELL (1938): *Nature* 141, 747.

BAER, M. and S. ALTMAN (1985): A catalytic RNA and its gene from Salmonella typhimurium. *Science* 228, 999-1002.

BAHADUR, K. (1954): *Nature* 173, 1141-1142.

BAHADUR, K. (1964): Conversion of lifeless matter into the living system. *Zentralbl. Bakteriol. Abt. 2* 118, 671-694.

BAHADUR, K. (1966): *Synthesis of Jeewanu, the Protocells.* M/S Ramnarain Lal Beni Presad, Allahabad, India.

BAHADUR, K. (1967): Synthesis of Jeewanu, the protocell. *Zentralbl. Bakteriol. Abt. 2.* 121, 291-319.

BAHADUR, K. (1973/1974): Photochemical formation of self-sustaining coacervates. *Proc. Indian Acad. Sci.* B 39, 455-467.

BAHADUR, K. and S. RANGANAYAKI (1970): The photochemical formation of self-sustaining coacervates. *J. Br. Interplanet. Soc.* 23, 813-829.

BAHADUR, K. and S. RANGANAYAKI (1983): Synthesis of Jeewanu, the protocells and study of their functional properties namely the ability to split water in sunlight and to fix molecular nitrogen and carbon dioxide. Lecture, *7th ISSOL Conference,* Mainz.

BAHADUR, K.; S. RANGANAYAKI and S. SANTAMARIA (1958): *Nature* 182, 1668-1670.

BAHADUR, K; et al. (1963): *Vijnana Parishad Anusandhan Patrika* 6, 63-117.

BAHADUR, K; et al. (1966): Synthesis of molecular associations with the properties of biological order. *Zentralbl. Bakteriol. Abt 2* 120, 740-752.

BAIRD, A. K; et al. (1976): Mineralogic and petrologic implications of Viking geochemical results from Mars: Interim report. *Science* 194, 1288-1293.

BALASUBRAMANIAN, R. (1982): Origin of life: A hypothesis for the origin of adaptor-mediated ordered synthesis of proteins and an explanation for the choice of terminating codons in the genetic code. *BioSystems* 15, 99-104.

BARGHOORN, E. S. and W. J. SCHÖPF (1966): Microorganism three billion years old from the Precambrian of South Africa. *Science* 152, 758-763.

BAROSS, J. A. and J. W. DEMING (1983): Growth of "black smoker" bacteria at temperatures of at least 250 °C. *Nature* 302, 423-426.

BAROSS, J. A.; M. D. LILLEY, and L.I. GORDON (1982): Is the CH_4, H_2, and CO venting from submarine hydrothermal systems produced by thermophilic bacteria? *Nature* 298, 366-368.

BARRICELLI, N. A. (1971): On the origin and evolution of the genetic code I. Wobbling and its potential significance. *J. Theor. Biol.* 67, 85-89.

BAUER, E. (1963): *Theoretical Biology* (in Hungarian). Akadémiai Kiadó, Budapest.

BAUER, E. S. (1982): *Theoretical Biology.* Reprint of the 1935 edition with a preface, a biographical and critical essay. Hung. Acad. Sci.-Acad. Sci. USSR, Budapest.

BEATTY, J. K. (1979): Voyager's encore performance. *Sky and Telescope* 58, 206-216.

BÉKÉS, F.; T. GANTI; M. HIDVÉGI and M. KORPÁDI (1980): Unpublished results.

BERETTA, E.; F. VETRANO; F. SOLIMANO and C. LAZZARI (1979): Some results about nonlinear chemical systems represented by trees and cycles. *Bull. Math. Biol.* 41, 641-664.

BERNAL, J. D. (1959a): The problem of stages in biopoesis. In: A. J. Oparin (ed.): *The Origin of Life on Earth.* Pergamon Press, London.

BERNAL, J. D. (1959b): The scale of structural units in biopoesis. In: A. J. Oparin (ed.): *The Origin of Life on Earth.* Pergamon Press, London.

BERNAL, J. D. (1967): *The Origin of Life.* Weindenfeld and Nicolson, London.

BERNARD, C. (1878): *Lecons sur les phenomenes de la vie.* Paris.

BERTALANFFY, L. (1952): *Problems of Life. An Evolution of Modern Biological Thought.* Watts, London.

BIEMANN, K. (1974): Test results on the Viking gas chromatograph – Mars spectrometer experiments. *Orig. Life* 5, 417 -430.

BIEMANN, K. et al. (1977): The search for organic substances and inorganic volatile compounds in the surface of Mars. *J. Geophys*. Res. 82, 4641-4647.

BLANDINO, G. (1969): Theories on the nature of life. *Philosophical Library XIII.* New York.

BLITZ, L. (1982): Giant molecular-cloud complexes in the Galaxy. *Sci. Am.* 246, 72-80.

BLOOM, D. (1966): A speculation relating the genetic code and the evolutionary process. *Proc. Natl. Acad. Sci. USA* 55, 375-376.

BRASSEL, S. C; et al. (1981): Specific cyclic isoprenoids as biological markers of methanogenic bacteria in marine sediments. *Nature* 290, 693-696.

BRESCH, C.; U. NIESERT and D. HARNASCH (1980): Hypercycles, parasites and packages. *J. Theor. Biol.* 85, 399-405.

BRESLOW, R. (1959): On the mechanism of formose reaction. *Tetrahedron Lett.* 21, 22.

BRESLOW, R. et al. (1983): *J. Am. Chem. Soc.* 105, 2739.

BREWIN, N. (1972): Catalytic role for RNA in DNA replication. *Nature New Biol..* 236, 101.

BRIDGWATER, D; et al. (1981): Microfossil-like objects from the Archaean of Greenland: A cautionary note. *Nature* 289, 51-53.

BRIGGS, M. H. (1965): Experiments on the origin of cells. *Space Flight* 129-131.

BROOKS, J. and G. SHAW (1973): *Origin and Development of Living Systems.* Academic Press, London.

BUNGENBERG DE JONG, H. (1932): *Protoplasma* 15, 110.

BURLEY, F. W. (1965): Transphosphorylation of adenosine di- and tri-phosphates in the presence of calcium phosphate precipitates. *Nature* 208, 683-684.

BURTON, F. G.; R. LOHRMANN and L. E. ORGEL (1974): On the possible role of crystals in the origins of life. 7. The adsorption and polymerization of phosphoramidates by montmorillonite clay. *J. Mol. Evol.* 3, 141-150.

BUTLEROV, A. (1861): Bildung einer zuckerartigen Substanz durch Synthese. *Liebigs Ann. Chem.* 120, 295.

BUVET, R. (1974): *L'origine des etres* vivants *et des processus biologiques.* Masson, Paris.

BUVET, R. and L. Le PORT (1973): Non-enzymic origin of the metabolism. *Space Life Sci.* 4, 434-447.

BUVET, R. and C. PONNAMPERUMA (eds.) (1971): *Chemical Evolution and the Origin of Life.* North-Holland, Amsterdam.

CAIRN-SMITH, A. G. and G. L. WALKER (1974): Primitive metabolism. *BioSystems* 5, 173-186.

CAIRN-SMITH, A. G.; P. INGRAM and G. L. WALKER (1972): Formose production by minerals: I Possible relevance to the origin of life. *J. Theor. Biol.* 35, 601.

CALVIN, M. (1969): *Chemical Evolution.* Oxford University Press, New York.

CANUTO, V. M; et al. (1983): The young Sun and the atmosphere and photochemistry of the early

Earth. *Nature* 305, 281-286.

CASSEN, P.; R. T. REYNOLDS and S. J. PEALE (1979): Is there liquid water on Europa? *Geophys. Res. Lett.* 6, 731-734.

CHADHA, M. S. and A.S.U. COUGHULEY (1983): Synthesis of prebiotic molecules–Role of some carbonyl compounds in prebiotic chemistry. *4th ISSOL Conference,* Mainz.

CHANG, S.; K. LENNON and J. VOLLMER (1977): Abundance and isotopic composition of carbon in Precambrian rocks of Greenland. *2nd ISSOL Conference,* Kyoto.

CHAPPE, B.; P. ALBRECHT and W. MICHAELIS (1982): Polar lipids of Archaebacteria in sediments and petroleums. *Science* 217, 65-66.

CHARGAFF, E. (ed.) (1955-1960): *The Nucleic Acids,* Vols. I–III. Academic Press, New York.

CHERNAVSKII, D. S. and N. M. CHERNAVSKAYA (1975): Some theoretical aspects of the problem of life origin. *J. Theor. Biol.* 50, 13-23.

CHOUGHULEY, A. S. U; et al. (1978): A possible prebiotic synthetis of thymine: Uracyl–formaldehyde–formic acid reaction. In: H. Onoda (ed.), *Proceedings of the 2nd ISSOL Meeting.* Center for Academic Publs. Japan, Tokyo.

CLARK, R. N. and T. B. McCORD (1980): The Galilean satellites: New near-infrared spectral reflectance measurements. *Icarus* 41, 323-339.

COMITA, P. B. and R. B. GAGOSIAN (1983): Membrane lipid from deep-sea hydrothermal vent methanogen: A new macrocyclic glycerol diether. *Science* 222, 1329-1331.

CORLISS, J. B. (1984): Energy currents and the origin of life on Earth (in Hungarian). *Fiz. Sz.* 34, 345-351.

COYNE, L. M. (1983): Clay energetics in chemical evolution. *IVth ISSOL Meeting,* Mainz.

CRAM, D.J. (1983): Cited by T. H. Maugh II. *Science* 223, 154-156.

CRAMER, F. (1961): *Angew. Chem.* 73, 49.

CRAMER, F. (1966): The synthesis of oligo- and polynucleotides. *Angew. Chem. Int. Ed.* Eng. 5, 173-184.

CRICK, F. H. C. (1966): The genetic code – Yesterday, today and tomorrow. *Cold Sprlng Harbor Symp. Quant. Biol.* 31, 3-9.

CRICK, F. H. C. (1967): Origin of the genetic code. *Nature* 213, 119.

CRICK, F. H. C. (1968): The origin of the genetic code. *J. Mol. Biol.* 38, 367-379.

CROTHERS, D. M. (1982): Nucleic acid aggregation geometry and the possible evolutionary origin of ribosomes and the genetic code. *J. Mol. Biol.* 162, 379-391.

CSÁNYI, V. (1978): General theory of evolution (in Hungarian). *Fiz. Sz.* 28, 401-443.

CSÁNYI, V. (1980): The general theory of evolution. *Acta Biol. Hung. Acad. Sci.* 31, 409-434.

CSÁNYI, V. (1981): General theory of evolution. *Soc. Gen. Syst. Res.* 6, 73-95.

CSÁNYI, V. (1982): *General Theory of Evolution.* Akadémiai Kiadó, Budapest.

CSÁNYI, V. (1988): *Evolutive Sytems: The General Theory of Evolution* (in Hungarian). Gondolat, Budapest.

CSÁNYI, V. and Gy. KAMPIS (1984): Autogenesis: The evolution of the self-organizing systems (in Hungarian). *Biologia* 32, 3-23.

CSÁNYI, V. and G. KAMPIS (1985): Autogenesis: The evolution of replicative systems. *J. Theor. Biol.* 114, 303-321.

CSER, F.; GY. HARDY and Z. VEKSLI (1974): Investigations on the polymerization in solid phase. XXXIII. (in Hungarian). *Magy. Kem. Foly.* 80, 177-181.

LeDANTEC, F. (1900): *The Nature and Origin of Life.* Hodder and Stoughton, London and A. S. Barnes and Co., New York.

LeDANTEC, F. (1910): *The Elements of the Biological Philosophy* (in Hungarian). Atheneum, Budapest.

DAYHOFF, M. O. (1972): *Atlas of Protein Sequence and Structure.* National Biomedical Research Foundation, Washington.

DEAMER, D. W. and J. ORÓ (1980): Role of lipids in prebiotic structures. *BioSystems* 12, 167-175.

DECKER, P. (1973): Evolution in open systems: Bistability and the origin of molecular asymmetry. *Nature New Biol.* 241, 72.

DECKER, P. (1974): *Evolution in offenen Systemen.* Bayerischen Akademie der Wissenschaften, Hannover.

DECKER, P. (1978): Inverse assimilation: A general principle of evolution of planetary surfaces. In: H. Noda (ed.), *Proc.IInd ISSOL Meeting.* Center for Academic Publications, Japan, Tokyo, pp. 631-637.

DECKER, P. and W. HEIDMANN (1978): Evolution in open systems. Acquisition and conversion of information in bioids. In: H. Noda (ed.), *Proc. IInd ISSOL Meeting.* Center for Academic Publications, Japan, Tokyo, pp. 617-624.

DECKER, P. and H. SCHVEER (1983): Reactions in the formol bioid: The origin of the branched chains of isoprenoids, valine and leucines. *4th ISSOL Meeting,* Mainz.

DECKER, P. and A. SPEIDEL (1972): Open systems which can mutate between several steady states ("bioids") and a possible prebiological role of the autocatalytic condensation of formaldehyde. *Z. Naturforsch.* 276, 257-263.

DEGANI, Ch. and M. HALMANN (1967): Chemical evolution of carbohydrate metabolism. *Nature* 216, 1207.

DEGANI, Ch. and M. HALMANN (1971): Primeval phosphate transfer processes leading to sugar phosphates. In: *1st European Biophysics Congress.* Verlag Wiener Med. Akademie, Wien.

DEGANI, Ch. and M. HALMANN (1972): Synthesis of α-D-glucose-1,6-diphosphate in potentially prebiotic conditions. *Nature New Biol.* 235, 171-173.

DEGENS, E. T.; J. MATHEJA and T. A. JACKSON (1970): Template catalysis: Asymmetric polymerization of amino acids on clay minerals. *Nature* 227, 492-494.

DEMETRIUS, L. (1983): Selection and evolution in macromolecular systems. *J. Theor. Biol.* 103, 619-643.

DeMORE, W. B. and Y. L. YOUNG (1982): Catalytic processes in the atmospheres of Earth and Venus. *Science* 217, 1209-1213.

DÉNES, F. and S. W. FOX (1976): Polycondensation of α-amino acids by pyrosulfuric acid. *BioSystems* 8, 83-88.

DeROSA, M.; et al. (1980): *Phytochemistry* 19, 249.

DIX, D. (1983): Toward a definition of life: Semantic and thermodynamic considerations. *J. Theor. Biol.* 102, 337-340.

DONAHUE, T. M.; J. H. HOFFMAN; R. R. HODGES Jr. and A. J. WATSON (1982): Venus was wet: A measurement of the ratio of deuterium to hydrogen. *Science* 216, 630-633.

DONN, B. (1982): Comets: Chemistry and chemical evolution. *J. Mol. Evol.* 18, 157-160.

DOSE, K. (1974): Chemical and catalytical properties of thermal polymers of amino acids (proteinoids). *Orig. Life* 5, 239-252.

DOSE, K. and B. RAJEWSKY (1957): *Biochim. Biophys. Acta* 25, 225.

DOWD, P. and J. M. SHAPIRO (1976): A nonenzymic model intermediate for the coenzymic B_{12} dependent isomerization of methylmalonyl coenzyme A to succinic coenzyme A. *J. Am. Chem. Soc.* 98, 3724-3725.

DOWLER, M. J.; W. D. FULLER and L. E. ORGEL (1970): Prebiotic synthesis of propionaldehyde and nicotinamide. *Science* 169, 1320-1321.

DRAGANIC, I. G. and Z. D. DRAGANIC (1971): *The Radiation Chemistry of Water.* Academic Press, New York.

DRAGANIC, I. G. and Z. D. DRAGANIC (1978): Radiation-chemical aspects of primordial organic synthesis on the primitive Earth and in extraterrestrial physicochemical processes. *XXIXth International Astronautical Congress,* Dubrovnik.

DRAGANIC, Z. D.; I. G. DRAGANIC and V. NIKETIC (1977): The radiation chemistry of aqueous solutions of ammonium and sodium cyanides in the megarad dose range. *Radiat. Res.* 69, 223-229.

EAKIN, R. E. (1963): An approach to the evolution of metabolism. *Proc. Natl. Acad. Sci. USA* 49, 360-368.

EIGEN, M. (1971): Self-organization of matter and the evolution of biological macromolecules.

Naturwissenschaften 58, 465-523.

EIGEN, M. and P. SCHUSTER (1977): The hypercycle. A principle of natural self-organization. Part A: Emergence of a hypercycle. *Naturwissenschaften* 4, 541-565.

EIGEN, M. and P. SCHUSTER (1978a): The hypercycle. A principle of natural self-organization. Part B: The abstract hypercycle. *Naturwissenschaften* 65, 7-41.

EIGEN, M. and P. SCHUSTER (1978b): The hypercycle. A principle of natural self-organization. Part C: The realistic hypercycle. *Naturwissenschaften* 5, 341-369.

EIGEN, M. and P. SCHUSTER (1979): *A Principle of Natural Self Organization.* Springer-Verlag, Berlin.

EIGEN, M. and P. SCHUSTER (1981): Comments on "Growth of a hypercycle" by King. *BioSystems* 13, 235.

EIGEN, M. and P. SCHUSTER (1982): Stages of life – Five principles of early organization. *J. Mol. Evol.* 19, 47-61.

EIGEN, M. and R. WINKLER-OSWATITSCH (1981a): Transfer-RNA: The early adaptor. *Naturwissenschaften* 68, 217-228.

EIGEN, M. and R. WINKLER-OSWATITSCH (1981b): Transfer-RNA, an early gene? *Naturwissenschaften* 68, 282-292.

EIGEN, M.; W. C. GARDINER Jr. and P. SCHUSTER (1980): Hypercycles and compartments. *J. Theor. Biol.* 85, 407-411.

EIGEN, M.; et al. (1980): Elementary step dynamics of catalytic hypercycles. *BioSystems* 12, 1-22.

EIGEN, M.; et al. (1981): The origin of genetic information. *Sci. Am.* 244, 78-94.

ENGELS, F. (1963): *Dialectics of Nature* (in Hungarian). Kossuth, Budapest.

ESHELMAN, R.; G. F. LINDAL and G. L. TYLER (1983): Is Titan wet or dry? *Science* 221, 53-55.

Exotic Biochemistry in Exobiology. (1970): Manuscript by unknown author.

FAKHRAI, H.; J. H. G. von ROODE and L. E. ORGEL (1981): Synthesis of oligoguanylates on oligocytidylate templates. *J. Mol. Evol.* 17, 295-302.

FARMER, C. B.; D. W. DAVIES and A. L. HOLLAND (1977): Mars: Water vapour observations from the Viking orbiters. *J. Geophys. Res.* 82, 4225-4248.

FERRIS, J. P. (1968): Cyanovinylphosphate: A prebiological phosphorylating agent? *Science* 161, 53-54.

FERRIS, J. P. (1979): *Science* 203, 1135-1136.

FERRIS, J. P. and J. E. KUDER (1970): Chemical evolution. 3. The photochemical conversion of enaminonitriles to imidazoles. *J. Am. Chem. Soc.* 92, 2527-2533.

FERRIS, J. P. and D. E. NICODEM (1972): *Nature* 238, 268-269.

FERRIS, J. P.; R. A. SANCHEZ and L. E. ORGEL (1968): Studies in prebiotic synthesis. III.

Synthesis of pyrimidines from cyanoacetylene and cyanate. *J. Mol. Biol.* 33, 693-704.

FERRIS, J. P.; J. E. KUDER and A. W. CATALARO (1969): Photochemical reactions and the chemical evolution of purines and nicotinamide derivatives. *Science* 166, 765-766.

FERRIS, J. P.; G. GOLDSTEIN and D. J. HEAULIEN (1970): Chemical evolution. 4. An evolution of cyanovinyl phosphate as a prebiotic phosphorylating agent. *J. Am. Chem. Soc.* 92, 6598-6603.

FERRIS, J. P.; D. R. DONER and A. P. LOBO (1973): Possible role of HCN in chemical evolution. *J. Mol. Biol.* 74, 499-518.

FERRIS, J. P.; et al. (1974): Biomolecules from HCN. *Orig. Life* 153-157.

FERRIS, J. P.; O. S. ZAMEK; A. M. ALTBUCH and H. FREIMAN (1974): Chemical evolution. 13: Synthesis of pyrimidines from guanidine and cyanoacetaldehyde. *J. Mol. Evol.* 3. 301-309.

FERRIS, J. P.; P. C. JOSHI and J. G. LAWLESS (1978a): Pyrimidines from HCN. In: H. Noda (ed.): *Proceedings of the 2nd ISSOL Meeting.* Center for Academic Publications, Japan, Tokyo.

FERRIS, J. P.; P. C. JOSHI; E. H. EDELSON and J. G. LAWLESS (1978b): HCN: A plausible source of purines, pyrimidines and amino acids on the primitive Earth. *J. Mol. Evol.* 11, 293-311.

FLASAR, F. M. (1983): Oceans on Titan? *Science* 221, 55-57.

FLORES, J. I. and C. PONNAMPERUMA (1972): Polymerization of amino acids under primitive Earth conditions. *J. Mol. Evol.* 2, 1-9.

FOLSOME, C. E. (1976): Synthetic organic microstructures and the origins of cellular life. *Naturwissenschaften* 63, 303-307.

FOLSOME, C. (1977): Synthetic organic microstructures as model systems for early protobionts. In: C. Ponnamperuma (ed.), *Chemical Evolution of the Early Precambrian.* Academic Press, New York, pp. 171-179.

FOLSOME, C. and A. BRITTAIN (1981): Model protocells photochemically reduce carbonate to organic carbon. *Nature* 291, 482-484.

FOX, S. W. (1964): Thermal polymerization of amino acids and production of formed microparticles on lava. *Nature* 201, 336-337.

FOX, S. W. (1965): A theory of macromolecular and cellular origins. *Nature* 205, 328-340.

FOX, S. W. (1973a): Molecular evolution to the first cells. *Pure Appl. Chem.* 34, 641.

FOX, S. W. (1973b): The rapid evolution of complex systems from simple beginnings. In: M. Morois (ed.), *From Theoretical Physics to Biology.* Karger, Basel, pp. 133-144.

FOX, S. W. (1973c): Origin of the cell. Experiments and premises. *Naturwissenschaften* 60, 359-368.

FOX, S. W. (1974a): Origins of biological information and the genetic code. *Mol. Cell Biochem.* 3, 129.

FOX, S. W. (1974b): Coacervate droplets, proteinoid microspheres and the genetic apparatus. In: K. Dose et al. (eds.), *The Origin of Life and Evolutionary Biochemistry.* Plenum Press, New York, pp. 119-132.

FOX, S. W. (1975): Stereomolecular interactions and microsystems in experimental protobiogenesis. *BioSystems* 7, 22-36.

FOX, S. W. and K. DOSE (1972): *Molecular Evolution and the Origin of Life.* Freeman, San Francisco.

FOX, S. W. and K. HARADA (1960): The thermal copolymerization of amino acids common to proteins. *J. Am. Chem. Soc.* 82, 3745-3751.

FOX, S. W. and F. SUZUKI (1976): Linkages in thermal copolymers of lysine. *BioSystems* 8, 40-44.

FOX, S. W. and T. V. WAEHNELDT (1968): The thermal synthesis of neutral and basic proteinoids. *Biochim. Biophys. Acta* 160, 246-249.

FOX, S. W. and C. R. WINDSOR (1970): Synthesis of amino acids by heating of formaldehyde and ammonia. *Science* 170, 984.

FOX, S. W.; K. HARADA and A. VEGOTSKY (1959): Thermal polymerization of amino acids and a theory of biochemical origins. *Experientia* 15, 81-84.

FOX, S. W.; R. J. McCAULEY and A. WOOD (1967): A model of primitive heterotrophic proliferation. *Comp. Biochem. Physiol.* 20, 773.

FOX, S. W.; J. C. LACEY Jr. and T. NAKASHIMA (1971): Interactions of thermal proteinoids with polynucleotides. In: D. W. Ribbons and V. F. Woessuer (eds.), *Nucleic Acid–Protein Interactions.* North-Holland, Amsterdam, pp. 113-127.

FRANK, G. M. (1962): Remarks. In: *The Philosophical Problems of Modern Natural Sciences* (in Hungarian). Akadémiai Kiadó, Budapest.

FRASER, C. L. and C. E. FOLSOME (1975): Exponential kinetics of formulation of organic microstructures. *Orig. Life* 6, 429-433.

FREITAS, A. R., Jr. (1983): Naming extraterrestrial life. *Nature* 301, 106.

FRIEDMANN, N. and S. L. MILLER (1969): Phenylalanine and tyrosine synthesis under primitive Earth conditions. *Science* 166, 766-767.

FRIEDMANN, N.; S. L. MILLER and R. A. SANCHEZ (1971): Primitive Earth synthesis of nicotinic acid derivatives. *Science* 171, 1026-1027.

FURBERG, S. (1952): *Acta Chem. Scand.* 6, 634.

GABEL, N. W. and C. PONNAMPERUMA (1967): Model for the origin of monosaccharides. *Nature* 216, 453-454.

GAMOW, C. (1954): Possible relation between desoxyribonucleic acid and protein structures. *Nature* 173, 318.

GÁNTI, T. (1962): The structure of deoxyribonucleic acid (in Hungarian). *Magy. Kem. Lapja* 117, 289-296.

GÁNTI, T. (1966): *Revolution in the Life-research* (in Hungarian). Gondolat, Budapest.

GÁNTI, T. (1971): *The Principle of Life,* 1st ed. (in Hungarian). Gondolat, Budapest.

GÁNTI, T. (1972): Investigations on the organization of living systems (in Hungarian). *1. MTA Biol. Oszt. Kozl.* I5, 177-196.

GÁNTI, T. (1974a): The origin of life – Recent investigations and results. In: Gy. Csaba (ed.), *The Actual Problems of Biology,* Vol. 1 (in Hungarian). Medicina, Budapest, pp. 45-90.

GÁNTI, T. (1974b): Theoretical deduction of the structure and function of the genetic substance (in Hungarian). *Biologia* 22, 17-35.

GÁNTI, T. (1975): Synthesis by the enzyme systems of yeast (in Hungarian). *Szeszipar* 23, 57-63.

GÁNTI, T. (1977): Life criteria and the definition of living systems (in Hungarian). *Biologia* 25, 91-107.

GÁNTI, T. (1978a): Self-regulation and self-organization of molecular supersystems. In: Gy. Csaba (ed.), *Biological regulation* (in Hungarian). Medicina, Budapest, pp. 37-73.

GÁNTI, T. (1978b): *The Principle of Life* 2nd revised ed. (in Hungarian). Gondolat, Budapest. 6th ed. (in English) 1988. OMIKK, Budapest.

GÁNTI, T. (1978c): On the reality of extraterrestrial biogenesis. *XXIXth IAF Congress,* Dubrovnik, Preprint No. IAF-78-A-51.

GÁNTI, T. (1978d): Viking biological experiments (in Hungarian). *Fiz. Sz.* 28, 371-376.

GÁNTI, T. (1979a): *A Theory of Biochemical Supersystems and its Application to Problems of Natural and Artificial Biogenesis.* Akadémiai Kiadó, Budapest and University Park Press, Baltimore.

GÁNTI, T. (1979b): The paradox of individuality (in Hungarian). *Biologia* 27, 100-102.

GÁNTI, T. (1979c): Interpretation of the prebiotic evolution on the basis of the chemoton theory (in Hungarian). *Biologia* 27, 161-175.

GÁNTI, T. (1980): *The Life and its Origin* (in Hungarian). Tankönyvkiadó, Budapest. Russian edition: Zhizny i io proishozhgyenii. 1984. Prosveshchenie, Moscow.

GÁNTI, T. (1983): The origin of the earliest sequences (in Hungarian). *Biologia* 31, 47-54.

GÁNTI, T. and I. ALMÁR (1982): Earth-independent human colonies on the Galilean moons of Jupiter. *Earth-Oriented Appl. Space Technol.* 2, 63-67.

GÁNTI, T. and Á. CSÓKA (1975): Production of phosphoenol-pyruvic acid from 3-phosphoglyceric acid (in Hungarian). *Magy. Kem. Foly.* 81, 335-336.

GÁNTI, T. and Cs. GÁSPÁR (1978): The conditions for a growing microsphere to divide. *Acta Chim. Acad. Sci. Hung.* 98, 278-283.

GÁNTI, T.; L. PATAI, R. BODOKI and T. PALÁGYI (1970): Hung. Pat. No. 160180.

GARRISON, W. M.; et al. (1951): The reduction of carbon dioxide in aqueous solutions by ionizing radiation. *Science* 114, 461.

GEORGOPAPADOKOU, N. H. and A. I. SCOTT (1977): On B_{12} biosynthesis and evolution. *J. Theor. Biol.* 69, 381-384.

GIBBS, D.; R. LOHRMANN and L. E. ORGEL (1980): Template-directed synthesis and selective adsorption of oligoadenylates on hydroxyapatite. *J. Mol. Evol.* 15, 347-354.

GLUCHKOV, V. M.; V. V. IVANOV and V. M. YANENKO (1970): *Modelirovniye Rasvivayushchichsya Sistem.* Nauka, Moscow.

GOLDANSKI, V. I. (1977): Mechanism for formaldehyde polymer formation in interstellar space. *Nature* 268, 312-313.

GRAFSTEIN, D. (1983): Stereochemical origins of the genetic code. *J. Theor. Biol.* 105, 157-174.

GRAY, W. (1975): Emotional cognitive structure theory and the development of a general systems psychotherapy. *Gen. Syst.* 20, 17-23.

GREENBERG, J. M. (1984): The structure and evolution of interstellar grains. *Sci. Am.* 250, 96-107.

GROSSENBACHER, K. A. and C. A. KNIGHT (1965): In S. W. Fox (ed.), *Origins of Prebiological Systems and of their Molecular Matrices.* Academic Press, New York, p. 173.

GROVES, D. I.; J. S. R. DUNLOP and R. BUICK (1981): An early habitat of life. *Sci. Am.* 245, 56-65.

GUERRIER-TAKADA, C. and S. ALTMAN (1984): Catalytic activity of an RNA molecule prepared by transcription in vitro. *Science* 223, 285-286.

GUERRIER-TAKADA, C.; K. GARDINER; T. MARSH; N. PACE and S. ALTMAN (1983): *Cell* 19, 849.

GYARMATI, I. (1960): *Introduction to the Irreversible Thermodynamics* (in Hungarian). Mérnöktovábbképző Intézet, Budapest.

GYARMATI, I. (1970): *Nonequilibrium Thermodynamics.* Springer, Berlin.

HAGAN, W. I., Jr. and J. P. FERRIS (1983): Clay-mediated condensation reactions of nucleotides. *IVth ISSOL Meeting,* Mainz.

HALDANE,J. B. S. (1954): The origins of life. *Nature New Biol.* 16, 12.

HALL, T. S. (1969): *Ideas of Life and Matter. Studies in the History of General Physiology 600 B.C. – 1900 A.D.,* Vols. 1 and 2. University of Chicago Press, Chicago.

HALMANN, M. (1975): Models of prebiological phosphorylation. *Orig. Life* 6, 169-174.

HALMANN, M. and S. BLOCH (1979): Glyoxal and malonaldehyde formation by ultraviolet irradiation of aqueous formaldehyde. *BioSystems* 11, 227-232.

HALMANN, M.; R. A. SANCHEZ and L. E. ORGEL (1969): Phosphorylation of D-ribose in aqueous solution. *J. Org. Chem.* 34, 3702-3703.

HARADA, K. and S. W. FOX (1964): Thermal synthesis of natural amino acids from a postulated primitive terrestrial atmosphere. *Nature* 201, 335-336.

HARADA, K. and S. SUZUKI (1977): Formation of amino acids from elemental carbon by contact glow discharge electrolysis. *Nature* 266, 275-276.

HARDENBECK, H. G.; G. KRAMPITZ and L. WULF (1968): Decarboxylation of pyruvic acid in aqueous solution by thermal proteinoids. *Arch. Biochem.* 123, 72-81.

HART, M. H. (1978): The evolution of the atmosphere of the Earth. *Icarus* 33, 23-39.

HART, M. H. (1979): Was the pre-biotic atmosphere of the Earth heavily reducing? *Orig. Life* 9, 261-266.

HARTMANN, I.; T. NAWROTH and K. DOSE (1983): Formation of specific amino acid sequences during carbodiimide-mediated condensation of amino acids in aqueous solution, and computer simulated sequence generation. *4th ISSOL Meeting,* Mainz.

HARTMANN, M. (1925): *Allgemeine Biologie.* Jena.

HARVEY, S. C. and A. McCAMMON (1981): Intramolecular flexibility in phenylalanine transfer RNA. *Nature* 294, 286-287.

HASSELSTORM, T.; M. C. HENRY and B. MURR (1975): *Science* 125, 350.

HATTORI, Y. (1983): Formation of amino acids from CH_4-rich or CO_2-rich model atmosphere. *Lecture in the 4th ISSOL Conference,* Mainz.

HAYATSU, R.; M. H. STUDIER; S. MATSUOKA and E. ANDERS (1972): Origin of organic matter in early solar system. 6. Catalytic synthesis of nitriles, nitrogen bases and porphyrin-like pigments. *Geochim. Cosmochim. Acta* 36, 555-571.

HEIDMANN, W.; P. DECKER and R. POHLMANN (1978): Cooperative kinetics in the formaldehyde bioid: An autocatalysis second order in products. In: H. Noda (ed.), *Proc. 11th ISSOL Meeting.* Center for Academic Publications, Japan, Tokyo, pp. 625-630.

HEINRICH, R.; S. M. RAPOPORT and T. A. RAPOPORT (1977): Metabolic regulation and mathematical models. *Prog. Biophys. Mol. Biol.* 32, 1-82.

HENDRY, L. D. and E. D. BRANSONE, Jr. (1981): Are there structural analogies between amino acids and nucleic acids? *Orig. Life* 11, 203-211.

HENDRY, L. B. and F. H. WITHAM (1979): Stereochemical recognition in nucleic acid–amino acid interactions. Its implication in biological coding: A model approach. *Perspect. Biol. Med.* 22, 333-345.

HERRERA, A. (1928): *Colloid Chem.*2, 81.

HERRERA, A. (1934): *Bull. Labor plasmogénie* Mexico 1, 49; 1, 63.

HERRERA, A. (1942): *Science* 96, *14.*

HESS, S. L.; et al. (1976a): Mars climatology from Viking 1 after 20 sols. *Science* 194, 78-81.

HESS, S. L.; et al. (1976b): Early meteorological results from the Viking 2 lander. *Science* 194, 1352-

1353.

HESS, S. L.; et al. (1977): Meteorological results from the surface of Mars: Viking 1 and 2. *J. Geophys. Res.* 82, 4559-4574.

HIDVÉGI, M. (1980): Manuscript.

HODGSON, G. W. and B. L. BAKER (1967): Porphyrin abiogenesis from pyrrole and formaldehyde under simulated geochemical conditions. *Nature* 216, 29-32.

HODGSON, G. W. and C. PONNAMPERUMA (1968): Prebiotic porphyrin genesis: Porphyrin from electric discharge in methane, ammonia and water vapor. *Proc. Natl. Acad. Sci. USA* 59, 22-28.

HOFFMANN, G. W. (1975): Stochastic theory of the origin of the genetic code. *Annu. Rev. Phys. Chem.* 26, 123-144.

HOROWITZ, N. H. and G. L. HOBBY (1977): Viking on Mars: The carbon assimilation experiments. *J. Geophys. Res.* 82, 4659-4662.

HOROWITZ, N. H.; G. L. HOBBY and J. S. HUBBARD (1976): The Viking carbon assimilation experiments: Interim report. *Science* 174, 1321-1322.

HORTOBÁGYI, T. and J. VIGASSY (1966): Microorganisms in the water circuits exposed to radiation of the nuclear reactor Budapest-Csillebérc (in Hungarian). *KFKI Kozl.* 14, 235-243.

HORTOBÁGYI, T. and J. VIGASSY (1967): Microorganisms in the water circuits exposed to radiation of the nuclear reactor Budapest-Csillebérc. *Acta Biol. Hung.* 16, 151-160.

HSU, L. L.; S. BROKE and S. W. FOX (1971): Conjugation of proteinoid microspheres: A model of primordial communication. *Curr. Mod. Biol.*4, 12.

HUSSMANN, E. (1968): Zur Bildung von stickstoffhaltigen Heterocyclen nach Röntgenbestrahlung von wässrigen Lösungen aliphatischer Amine und Aminosäuren. 2. Arbeitstagung über extraterr. Biophys. Biol. 1967 Marburg. Tagungsbericht. Red: H. Bücker 225-233.

HUZELLA, T. (1933): *General Biology* (in Hungarian). Magyar Orvosi Könyvkiadó Vállalat, Budapest.

ISHIGAMI, M. and K. NAGANO (1975): The origin of the genetic code. *Orig. Life* 6, 551-560.

ISHIGAMI, M.; K. NAGANO and N. TANATSUKA (1977): The origin of the protein synthesis mechanism. *BioSystems* 9, 229-243.

ISHIMA, Y.; A. T. PRZYBYLSKI and S. W. FOX (1981): Electrical membrane phenomena in spherules from proteinoid and lecithin. *BioSystems* 13, 243-251.

IUB International Union of Biochemistry (1979): *Enzyme Nomenclature 1978.* Academic Press, New York.

JACOB, F. (1970): *La logique du vivant. Une histoire de l'hérédité.* Éditions Gallimard, Paris.

JOHNSON, T. V. and L. A. SODERBLOM (1983): lo. *Sci. Am.* 249, 60-71.

JOYCE, G. F. (1983a): The instability of the autogen. *J. Mol. Evol.* 19, 192-194.

JOYCE, G. F. (1983b): Non-enzymatic template-directed synthesis of RNA copolymers. *4th ISSOL Meeting,* Mainz.

JOYCE, G. F, et al. (1984): Chiral selection in poly(C)-directed synthesis of oligo(G). *Nature* 310, 602-604.

JUKES, T. H. (1965): Coding triplets and their possible evolutionary implications. *Biochem. Biophys. Res. Commun.* 19, 391-394.

JUKES, T. H. (1974): On the possible origin and evolution of the genetic code. *Orig.Life* 5, 331-350.

JUKES, T. H. (1981): Amino acid codes in mitochondria as possible clues to primitive codes. *J. Mol. Evol.* 18, 15-17.

JUKES, T. H. (1983): Evolution of the amino acid code inferences from mitochondrial codes. *J. Mol. Evol.* 19, 219-225.

JUNGCK, I. R. (1978): The genetic code as a periodic table. *J. Mol. Evol..* 11, 211-224.

JURKA, J.; Z. KOLOSZA and I. ROTERMAN (1982): Globular protein, G. U. wobbling, and the evolution of the genetic code. *J. Mol. Evol.* 19, 20-27.

KAHANE, E. (1962): *La Vie n'existe pas!* Les Editions de l'Union Rationaliste, Paris.

KAISER, E. T. and D. S. LAWRENCE (1984): Chemical mutation of enzyme active sites. *Science* 226, 505-511.

KENYON, D. H. and A. NISSENBAUM (1976): Melanoidin and aldocyanin microspheres: Implications for chemical evolution and early Precambrian micropaleontology. *J. Mol. Evol.* 7, 245-251.

KEOSIAN, J. (1968): *Origin of Life.* Reinhold, New York.

KHORANA, H. G. (1961): *Some Recent Developments in the Chemistry of Phosphate Esters of Biological Interest.* Wiley, New York.

KHORANA, H. G. and A. R. TODD (1953): *J. Chem. Soc. London* 2257.

KIEFFER, H. H.; et al. (1977): Thermal and albedo mapping of Mars during the Viking primary mission. *J. Geophys. Res.* 82, 4249-4291.

KING, G. A. M. (1981): Growth of a hypercycle and comparison with conventional autocatalysis. *BioSystems* 13, 225-234.

KING, G. A. M. (1982): Recycling, reproduction, and life's origins. *BioSystems* 15, 89-97.

KLEIN, H. P.; et al. (1916): The Viking biological investigation: Preliminary results. *Science* 194, 99-105.

KNOBEL, R.; F. FREUND; H. BREUER; M. SCHÖDER and G. NITZ (1983): Formation of organic molecules from carbon, hydrogen and nitrogen codissolved in minerals. *IVth ISSOL Meeting,* Mainz.

KNOLL, A. H. and E. S. BARGHOORN (1977): Archean microfossils showing cell division from the Swaziland system of South Africa. *Science* 198, 196-198.

KORÁNYI, P. (1982a): The origin and early evolution of the enzymes and the appearance of these of the metabolism of the presently living systems (in Hungarian). Dissertation, ELTE. Budapest.

KORÁNYI, P. (1982b): Interpretation of possible enzymatic function of t-RNA-s on the basis of their general three-dimensional structure (in Hungarian). *Biologia* 30, 3-9.

KORÁNYI, P. and T. GÁNTI (1981): Coenzymes as remnants of ancient enzyme ribonucleic acid molecules (in Hungarian). *Biologia* 29, 107-124.

KRANPRITZ, L.·O. (1969): Catalytic functions of thiamine diphosphate. *Annu. Rev. Biochem.* 38, 213-240.

KUHN, H. (1972): Self-organization of molecular systems and evolution of the genetic apparatus. *Angew. Chem. Int. Ed.* 11, 798-820.

KUHN, H. (1976): Model consideration for the origin of life. *Naturwissenschaften* 63, 68-80.

KUHN, H. and C. KUHN (1978): Evolution of a genetic code simulated with the computer. *Orig. Life* 9, 137-150.

KUHN, H. and J. WASER (1981): Molecular self-organization and the origin of life. *Angew. Chem. Int. Ed.* 20, 500-520.

KUHN, W. R. and S. K. ATREYA (1979): Ammonia photolysis and the greenhouse effect in the primordial atmosphere of the Earth. *Icarus* 37, 207-213.

KUHN, W. R. and J. F. KASTING (1983): Effects of increased CO_2 concentrations on surface temperature of the early Earth. *Nature* 301, 53-55.

KUMAR, S. (1976): Abiogenesis of nucleic acid bases. Ph. D. Thesis. Chemistry Department, Allahabad University, India.

KÜPPERS, D. (1975): The general principles of selection and evolution at the molecular level. *Prog. Biophys. Mol. Biol.* 30, 1-22.

KÜPPERS, D. (1979): Towards an experimental analysis of molecular self-organization and precellular Darwinian evolution. *Naturwissenschaften* 66, 228-243.

KUSHWAHA, S. C.; et al. (1981): *Science* 211, 1163.

LACEY, J. C., Jr.; A. L. WEBER and W. E. WHITE (1975): A model for the coevolution of the genetic code and the process of protein synthesis: Review and assessment. *Orig. Life* 6, 273-283.

LAHAV, N. (1975): Simultaneous existence of different environments in aqueous clay systems and its possible role in prebiotic synthesis. *J. Mol. Evol.* 5, 243-247.

LANGENBECK, W. (1942): Die Formaldehydkondenzation als organische Autokatalyse. *Naturwissenschaften* 30, 30.

LANGWORTHY, T. A.; T. G. TORNABENE and G. HOLZER (1982): *Zentralbl. Bakteriol. Parasitenkd. Infektionskr. Hyg. Abt. 1 Orig. C* 3, 228.

LANGWORTHY, T.; A et al. (1983): *Syst. Appl. MIcrobiol.* 4, 1.

LASAGA, A. C.; H. D. HOLLAND and M. T. DWYER (1971): Primordial oil slick. *Science* 174, 53-55.

LAVRENTIEV, G. A.; T. F. STRIGUNKOVA and I. A. EGAROV (1983): Abiological formation of organic compounds in conditions simulated volcanic ashes-gas cloud. *IVth ISSOL Meeting,* Mainz.

LEDUC, S. (1907): *Les bases physiques de la vie.* Paris.

LEDUC, S. (1910): *Theorie physicochimique de la vie.* Paris.

LEHMANN, U. and H. KUHN (1984): Origin of the code. *25th COSPAR Meeting.* Graz.

LEHN, J. M. (1985): Supramolecular chemistry: Receptors, catalysts and carriers. *Science* 227, 849-856.

LEIBNIZ, G. W. (1705): *Considerations sur les principes de vie et sur les natures plastiques par l auteur de systéme de l'harmonie préétablie.* Ed. Erdmann 429-432. Ed. Gerhardt VI. 539-546.

LEIBNIZ, G. W. (1710): *Epistola ad Wagnerum de vi activa corporis, de anima. de anima brutorum.* Ed. Erdmann 465-467. Ed. Gerhardt VII. 528-532.

LEIBNIZ, G. W. (1714): *La Monadologie.* Ed. Erdmann 705-712. Ed. Gerhardt VII. 607-625.

LENGYEL, S. and I. GYARMATI (1981a): Nonlinear thermodynamical studies of homogeneous chemical kinetic systems. *Period. Polytech. Chem. Eng.* 61, 104.

LENGYEL, S. and I. GYARMATI (1981b): On the thermodynamics of elementary chemical reactions in homogeneous systems. *J. Chem. Phys.* 75, 2384-2389.

LENGYEL, S. and I. GYARMATI (1986): Constitutive equations and reciprocal relations of non-ideal homogenous closed chemical kinetic systems. *Acta Chim. Hung.* 122, 7-17.

LEVINE, J. S. (1982): The photochemistry of the paleoatmosphere. *J. Mol. Evol.* 18, 161-172.

LEVINE, J. S.; T. R. AUGUSTSSON and M. NATARAIAN (1982): The prebiological paleoatmosphere: Stability and composition. *Orig. Life* 12, 245-259.

LEWIN, G. V. and P. A. STROAT (1976): Viking labeled release biology experiment: Interim results. *Science* 194, 1322-1329.

LEWIN, G. V. and P. A. STROAT (1977): Recent results from the Viking labeled release experiment on Mars. *J. Geophys. Res.* 82, 4663-4667.

LEWIN, R. (1984): First true RNA catalyst found. *Science* 223, 266-267.

LIEBIG, J. (1842): *Die Tierchemie oder organische Chemie in Ihrer Anwendung auf Agrikultur und Physiologie.*

LOHRMANN, R. (1972): Formation of urea and guanidine by irradiation of ammonium cyanide. *J. Mol. Evol.* 1, 263-269.

LOHRMANN, R. (1982): Oligonucleotide formation catalyzed by mononucleotide matrices. *J. Mol. Evol.* 18, 185-195.

LOHRMANN, R. and L. E. ORGEL (1971): Urea–inorganic phosphate mixtures as prebiotic phosphorylating agents. *Science* 171, 490-494.

LOHRMANN, R.; et al. (1975): Prebiotic peptide-formation in the solid state. 1. Reactions of benzoate ion and glycine with adenosine-5'-phosphorimidazolide. *J. Mol. Evol.* 5, 57-73.

LOHRMANN, R. and L. E. ORGEL (1977): Template-directed reactions of aminonucleosides. *J. Mol. Evol.* 9, 323-328.

LOHRMANN, R. and L. E. ORGEL (1978): Formation of P_1P_2-dinucleoside-5'-pyrophosphates under prebiological conditions. *J. Mol. Evol.* 11, 17-23.

LOHRMANN, R.; P. K. BRIDSON and L. E. ORGEL (1981): Condensation of activated diguanylates on a poly (C) template. *J. Mol. Evol.* 17, 303-306.

LOWE, D. Z. (1980): Stromatolites 3,400 Myr old from the Archean of Western Australia. *Nature* 284, 441- 443.

LOWE, C. U.; M. W. REES and R. MARKHAM (1963): Synthesis of complex organic compounds from simple precursors: Formation of amino acids, amino acid polymers, fatty acids and purines from ammonium cyanide. *Nature* 199, 219-222.

LÖWENSTEIN, J. M. (1958): Transphosphorylation catalyzed by bivalent metal ions. *Biochem. J.* 70, 222-230.

LUNINE, J. I.; D. I. STEVENSON and Y. L. YOUNG (1983): Ethane ocean on Titan. *Science* 222, 1229-1230.

LWOFF, A. (1962): *Biological Order.* MIT, Cambridge, Mass.

MACCHIATO, M. F. and A. TRAMONTANO (1982): Thermodynamic approach to a possible theory of the evolution of a genetic code. *Z. Naturforsch.* 37c, 1031-1037.

McALHANEY, W. W. and D. L. ROHLFING (1976): Formation of proteinoid microspheres under simulated prebiotic atmospheres under individual gases. *BioSystems* 8 ,45-50.

McELROY, M. B.; M. J. PRATHER and J. M. RODRIGUEZ (1982): Escape of hydrogen from Venus. *Science* 215, 1614-1615.

McGOVERN, W. E. (1969): The primitive Earth: Thermal models of the upper atmosphere for a methane-dominated environment. *J. Atmos. Sci.* 26, 623-635.

McKUSICK, B. C.; W. E. MOCHEL and F. W. STACEY (1960): *J. Am. Chem. Soc.* 82, 723.

MAKULA, R. A. and M. E. SINGER (1978): *Biochem. Biophys. Res. Commun..* 82, 716.

MANDEVILLE, J. C.; J. E. GEAKE and A. DOLLFUS (1980): Reflectance polarimetry of Callisto and the evolution of the Galilean satellites. *Icarus* 41, 343-355.

MANDL, I. and C. NEUBERG (1952): Preparation of **D**(−)-3-phosphoglyceric acid. *Metods Enzymol.* 111, 208.

MANN, A. P. C. and D. A. WILLIAMS (1980): A list of interstellar molecules. *Nature* 283, 721-725.

MARGULIS, L. (1970): *Origin of Eukaryotic Cells.* Yale University Press, New Haven, Conn.

MARQUAND, J. (1968): *Life: Its Nature, Origins and Distribution.* Oliver & Boyd, Edinburgh.

MASURSKY, H. and N. L. CRABILL (1976): Search for the Viking 2 landing site. *Science* 194, 62-68.

MASURSKY, H.; et al. (1977): Classification and time of formation of Martian channels based on Viking data. *J. Geophys. Res.* 82, 4016-4038.

MATSUNO, K. (1981): Material self-assembly as a physicochemical process. *BioSystems* 13, 237-241.

MATSUNO, K. (1982): Natural self-organization of polynucleotides in protobiogenesis: Appearance of a protohypercycle. *BioSystems* 15, 1-11.

MATTHEWS, C. N. (1975): The origin of proteins: Heteropolypeptides from hydrogen cyanide and water. *Orig. Life* 6, 155-162.

MATTHEWS, C. N. and R. E. MOSER (1966): Prebiological protein synthesis. *Proc. Natl. Acad. Sci. USA* 56, 1087-1094.

MATTHEWS, C. N. and E. R. MOSER (1967): Peptide synthesis from hydrogen cyanide and water. *Nature* 215, 1230-1235.

MATTHEWS, C. N.; et al. (1977): Deuterolysis of amino precursors: Evidence for hydrogen cyanide polymers as protein ancestors. *Science* 189, 622-625.

MATTHEWS, C. N.; R. LUDICKY and R. WIDING (1983): Heteropolypeptides from hydrogen cyanide and water? *4th ISSOL Meeting,* Mainz.

MATTHEWS, H. E. (1984): A new interstellar molecule: Tricarbon monoxide. *Nature* 310, 125-126.

MATURANA, H. R. (1981): Autopoesis. In: M. Zeleny (ed.), *Autopoesis.* North-Holland, Amsterdam.

MATURANA, H. R. and F. R. VARELA (1980): Autopoesis and cognition. *Boston Studies in Philosophy of Science,* Vol. 42. Reidel, Dordrecht.

MAUGH, T. H., II (1984): Need a catalyst? Design an enzyme. *Science* 223, 269-271.

METZLER, D. E.; M. IKOVA and E. E. SNELL (1954): A general mechanism for B_6-catalysed reactions. *J. Am. Chem. Soc.* 76, 648-652.

MICHELSON, A. M. and C. MONNY (1967): Polynucleotides X. Oligonucleotides and their association with polynucleotides. *Biochim. Biophys. Acta* 149, 1007-1026.

MIKELSAAR, H. N. (1975): A concept of amino acid archeocorrelation: Origin of life and the genetic code. *J. Theor. Biol.* 50, 203-212.

MILLER, J. G. (1965a): Living systems: Basic concept. *Behav. Sci.* 10, 193-237.

MILLER, J. G. (1965b): Living systems: Structure and process. *Behav. Sci.* 10, 337-339.

MILLER, J. G. (1965c): Living systems: Cross-level hypotheses. *Behav. Sci.* 10, 380-411.

MILLER, S. L. (1953): A production of amino acids under possible primitive Earth conditions.

Science 117, 528-529.

MILLER, S. L. (1955): Production of some organic compounds under possible primitive Earth conditions. *J. Am. Chem. Soc.* 77, 2351.

MILLER, S. L. (1957): The mechanism of synthesis of amino acids by electric discharges. *Biochim. Biophys. Acta* 23, 480-489.

MILLER, S. L. (1983): Prebiotic synthesis of amino acids, hydrogen cyanide, formaldehyde and ammonia in various primitive Earth atmospheres. *Lecture in the 4th ISSOL Conference,* Mainz.

MILLER, S. L. and H. C. UREY (1959): Organic compound synthesis on the primitive Earth. *Science* 120, 245-251.

MIZUHARA, S. and P. HANDLER (1954): Mechanism of thiamine-catalysed reactions. *J. Am. Chem. Soc.* 76, 571-573.

MIZUNO, T. and A. H. WEISS (1974): Synthesis and utilization of formose sugars. In: S. T. Tipson (ed.), *Advances in Carbohydrate Chemistry and Biochemistry.* Vol. 29. Academic Press, New York, pp. 173-227.

MIZUNO, T.; et al. (1970, 1971, 1972): Studies on synthesis and utilization of formose. *J. Agr. Chem. Soc. Japan.* 44, 324; 45, 344; 46, 73.

MIZUTANI, H.; et al. (1975): Study on the photochemical reactions of HCN. *Orig. Life* 6, 513-525.

MOISEEVA, L. N. (1971): A model of selective accumulation of carbohydrates diffusing thorugh artificial polymer membranes. In: R. Buvet and C. Ponnamperuma (eds.), *Chemical Evolution.* North-Holland, Amsterdam, pp. 363-368.

MOISEEVA, L. N.; T. E. PAVLOVSKAYA and R. F. SIRKU (1983): Abiogenic synthesis of peptides involving amino acyl adenylates in an open system. *4th ISSOL Meeting,* Mainz.

MONOD, J. (1970): *Le hasard et la necessité.* Editions du Senil, Paris.

MOORE, H. J.; et al. (1977): Surface materials of the Viking landing sites. *J. Geophys. Res.* 82, 4497-4523.

MORAVEK, J. and J. SKODA (1967): Thermic phosphorylations II. The formation of 3'-5' and 2'-5' dinucleoside phosphates through a reaction of uridine and 6-azauridine with inorganic phosphate. *Collect. Czech. Chem. Commun.* 32, 206-211.

MOROWITZ, H. and C. SAGAN (1967): Life in the clouds of Venus? *Nature* 215, 1259.

MOSER, R. E. and C. N. MATTHEWS (1968): Hydrolysis of aminoacetonitrile: Peptide formation. *Experientia* 15, 658-659.

MOSER, R. E.; A. R. CLAGETT and C. N. MATTHEWS (1968a): Peptide formation from aminomalononitrile (HCN trimer). *Tetrahedron Lett.* 13, 1605-1608.

MOSER, R. E.; A. R. CLAGETT and C. N. MATTHEWS (I968b): Peptide formation from diaminomaleonitrile (HCN tetramer). *Tetrahedron Lett.* 13, 1599-1603.

MUTCH, T. A.; et al. (1976): Fine particles on Mars: Observations with the Viking 1 lander cameras. *Science* 194, 87-91.

NAGY, B.; et al. (1981): Amino acids and hydrocarbons 3,800 Myr old in the Isua Rocks, southwestern Greenland. *Nature* 289, 53-56.

NAGY, L. A. and J. E. ZUMBERGE (1976): Fossil microorganisms from the approximately 2800 to 2500 million-year-old Hulawayan stromatolite: Application of ultramicrochemical analyses. *Proc. Natl. Acad. Sci. USA* 73, 2973-2976.

NAGYVÁRI, J. and J. H. FENDLER (1974): Origin of the genetic code: A physical-chemical model of primitive codon assignments. *Orig. Life* 5, 357-362.

NAKASHIMA, T. and S. W. FOX (1972): Selective condensation of amino-acyl adenylates by nucleoproteinoid microparticles. *Proc. Natl. Acad. Sci. USA* 69, 106- 108.

NELSESTUEN, G. L. (1980): Origin of life; considerations of alternatives to proteins and nucleic acids. *J. Mol. Evol.* 15, 59-72.

NEUBERG, C. H. and H. LUSTIG (1942):*J. Am. Chem. Soc.* 64, 2722.

NEUMANN, J. von (1948): General and logical theory of automata. *Lecture on the Hixon Symposion.* Pasadena, September 20.

NEUMANN, J. von (1966): *Theory of Self Reproducing Automata* (edited and completed by A. W. Burks). University of Illinois Press, Urbana.

NIKETIC, V.; et al. (1982): Enzymatic characterization of peptidic materials isolated from aqueous solutions of ammonium cyanide (pH 9) and hydrocyanic acid (pH 6) exposed ionizing radiation. *J. Mol. Evol.* 18, 130-136.

NIKETIC, V.; Z. DRAGANIC and I. DRAGANIC (1983): The radiolysis of aqueous cyanides: Polymers of interest to chemical evolution studies. *4th ISSOL Meeting,* Mainz.

NINIO, J. and L. E. ORGEL (1978): Heteropolynucleotides as templates for non-enzymatic polymerizations. *J. Mol. Evol.* 12, 91-99.

NISHIZAWA, M. and H. YANAGAWA (1983): A probable prebiotic peptide formation from glycineamide and related compounds. *4th ISSOL Meeting,* Mainz.

NOONER, D. W. and J. ORÓ (1974): Direct synthesis of polypeptides polycondensation of α-amino acids by polymetaphosphate esters. *J. Mol.Evol.* 3, 79-88.

OCHIAI, E. I. (1983): Inorganic aspects of chemical and biological evolution. *IVth ISSOL Meeting,* Mainz.

OHTA, T. and M. KIMURA (1971): Amino acid composition of proteins as a product of molecular evolution. *Science* 174, 170-253.

ONSAGER, L. (1931): Reciprocal relations in irreversible processes. I-II. *Phys. Rev.* 37, 405; 38, 2265.

OPARIN, A. J. (1957): *The Origin of Life on the Earth.* Oliver & Boyd, Edinburgh.

ORGEL, L. E. (1983): Template-directed synthesis. *4th ISSOL Meeting,* Mainz.

ORGEL, L. E. and R. LOHRMANN (1974): Prebiotic chemistry and nucleic acid replication. *Am. Chem. Res.* 7, 368-377.

ORÓ, J. (1960): Synthesis of adenine from ammonium cyanide. *Biochem. Biophys. Res. Commun.* 2, 407-412.

ORÓ, J. (1961): *Nature* 190, 389-390.

ORÓ, J. (1963): *Ann. N. Y. Acad. Sci.* 108, 464.

ORÓ, J. and A. P. KIMBALL (1961): Synthesis of purines under possible primitive Earth conditions. I. Adenine from hydrogen cyanide. *Arch. Biochem. Biophys.* 94, 217-227.

ORÓ, J. and A. P. KIMBALL (1962): Synthesis of purines under primitive Earth conditions. 2. Purine intermediates from hydrogen cyanide. *Arch. Biochem. Biophys.* 96, 293-313.

ORÓ, J. and E. STEPHEN-SHERWOOD (1974): The prebiotic synthesis of oligonucleotides. *Orig. Life* 5, 159-172.

ORÓ, J.; K. REWERS and D. ODOM (1982): Criteria for the emergence and evolution of life in the solar system. *Orig. Life* 12, 285-305.

ORÓ, J.; T. YAMROM; S. CORTES and B. BASILE (1983): The prebiotic synthesis and catalytic role of imidazoles and other condensing agents. *4th ISSOL Meeting,* Mainz.

OSHIMA, T. (1968): The catalytic hydrolysis of phosphate ester bonds by thermal polymers of amino acids. *Arch. Biochem.* 126, 478-485.

OSTER, G.; A. PERELSON and A. KATCHALSKY (1971): Network thermodynamics. *Nature* 234, 393-399.

OSTERBERG, R. and L. E. ORGEL (1972): Polyphosphate and trimetaphosphate formation under potentially prebiotic conditions. *J. Mol. Evol.* 1, 241-248.

OSTERBERG, R.; L. E. ORGEL and R. LOHRMANN (1973). Further studies on urea-catalyzed phosphorylation reaction. *J. Mol. Evol.* 2, 231-234.

OSTERN, P. and J. TERSZAKOWEC (1937): Über die enzymatische Synthese von Adenosin-5 -monophosphorsaure (Muskeladenylsaure) aus Adenosin. *Z. Physiol. Chem.* 250, 155-157.

OSTERN, P.; T. BARANOWSKY and J. TERSZAKOWEC (1938): Über die Phosphorylierung des Adenosins durch Hefe und die Bedeutung dieses Vergangs für die alkoholische Garung. II. *Z. Physiol. Chem.* 251, 258-284.

OSTERN, P.; J. TERSZAKOWEC and St. HUBL (1938): Über die Phosphorylierung des Adenosins durch Hefe und die Bedeutung dieses Vorgangs für die alkoholische Gärung. III. *Z. Physiol. Chem.* 255, 104-125.

OWEN, T. C. (1984a): Chemical evolution in the atmospheres of Jupiter and Titan. *XXVth COSPAR Meeting,* Graz.

OWEN, T. C. (1984b): The atmosphere and surface of Titan. *XXVth COSPAR Meeting,* Graz.

OWEN, T. C.; et al. (1977): The composition of the atmosphere at the surface of Mars. *J. Geophys. Res.* 82, 4635-4639.

OWEN, T. C.; R. D. CESS and V. RAMANATHAN (1979): *Nature* 277, 640-643.

OYAMA, V. I. and B. J. BERDAHL (1977): The Viking gas exchange experiment results from Chryse and Utopia surface samples. *J. Geophys. Res.* 82, 4669-4676.

PAECHT-HOROWITZ, M. (1974): The possible role of clays in prebiotic peptide synthesis. *Orig. Life* 5, 177-187.

PAECHT-HOROWITZ, M. (1983): Clays and minerals in prebiotic processes. *IVth ISSOL Meeting,* Mainz.

PAECHT-HOROWITZ, M. and N. LAHAV (1977): Polymerization of alanine in the presence of a non-swelling montmorillonite. *J. Mol. Evol.* 10, 73-76.

PAECHT-HOROWITZ, M.; I. BERGER and A. KATCHALSKY (1970): Prebiotic synthesis of polypeptides by heterogeneous polycondensation of amino-acid adenylates. *Nature* 228, 636-639.

PALM, C. and M. CALVIN (1962): Primordial organic chemistry. I. Compounds resulting from electron irradiation of $C^{14}H_4$. *J. Am. Chem. Soc.* 84, 2115-2121.

PATTEE, M. H. (1961): On the origin of macromolecular sequences. *Biophys. J.* 1, 683-710.

PELC, S. R. and M. G. E. WELTON (1966): Stereochemical relationships between coding triplets and amino acids. *Nature* 209, 868-870.

PFLUG, H. D. and H. JAESCHKE-BOYER (1978): Combined structural and chemical analysis of 3.800 Myr-old microfossils. *Nature* 280, 483-486.

POLLOCK, G. E. and R. HEIDERER (1979): Aldocyanine microspheres: Partial amino acid analysis of the microparticulates formed from simple reactants under various conditions. *J. Mol. Evol.* 13, 253-264.

PRIGOGINE, I. (1947): *Étude Thermodinamique des Phénoménes lreversibles.* DeSoer, Liege.

PRINN, R. G. (1985): The volcanoes and clouds of Venus. *Sci. Am.* 252, 36-43.

PRZYBYLSKI, A. T. and S. W. FOX (1984): Excitable artificial cells of proteinoid. *Appl. Biochem. Biotechnol.* 10, 301-307.

PRZYBYLSKI, A. T.; W. P. STRATTEN; R. M. SYREN and S. W. FOX (1982): Membrane, action, and oscillatory potentials in simulated protocells. *Naturwiss. Cushaften.* 69, 561-563.

QUAYLE, J. R. and T. FERENCI (1978): Evolutionary aspects of autotrophy. *Microbiol. Rev.* 42, 251 -273.

RABINOVITZ, J.; et al. (1969): Electric discharge reactions in mixtures of phosphine, methane, ammonia and water. *Nature* 224, 798.

RAINA, V. (1973): Studies in some aspect of abiogenesis of organic compounds of biological interests. *Ph. D. thesis,* Chemistry Department, Allahabad University, India.

RALPH, R. K. (1968): A suggestion on the origin of the genetic code. *Biochem. Biophys. Res. Commun.* 33, 213-218.

RANGANAYAKI, S.; V. RAINA and K. BAHADUR (1972): *J. Br. Interplanet. Soc.* 25, 279-285.

RAO, K. K.; M. W. W. ADAMS; P. MORRIS; D. O. HALL; S. RANGANAYAKI and K. BAHADUR (1980): Biophotolysis of water for H_2 production via natural and artificial catalytic systems. *Proc. Int. Symp. on Biological Applications of Solar Energy, Madurai, India, 1978.* The Macmillan Co. of India, Madras, pp. 201- 204.

RAO, M.; J. EICHBERG and J. ORÓ (1982): Synthesis of phosphatidylcholine under possible primitive Earth conditions. *J. Mol. Evol.* 18, 196-202.

RAULIN, F. (1984): Organic synthesis in gas phase and chemical evolution in planetary atmospheres. *XXVth COSPAR Meeting,* Graz.

RAULIN, F. and G. TOUPANCE (1975): Effect of H_2S on the formation of organic compounds from C, N, H, S model atmospheres submitted to a glow discharge. *Orig. Life* 507-512.

RAULIN, F.; D. MONREY and G. TOUPANCE (1982): Organic syntheses from CH_4–N_2 atmospheres: Implications for Titan. *Orig. Life* 12, 267-279.

REANNEY, D. C. (1977): Aminoacyl thiol esters and the origins of genetic specificity. *J. Theor. Biol.* 65, 555-569.

REICH, J. G. and E. E. SEL'KOV (1974): Mathematical analysis of metabolic networks. *FEBS Lett.* 40, 119-127.

REID, C. and L. E. ORGEL (1967): Synthesis of sugars in potentially prebiotic conditions. *Nature* 216, 455.

REINER,J. M. (1969): *Behavior of Enzyme Systems,* 2nd ed. Van Nostrand Reinhold, New York.

RENZ, M.; R. LOHRMANN and L. E. ORGEL (1971): Catalysts for the polymerization of adenosine cyclic 2', 3'-phosphate on a poly(U) template. *Biochim. Biophys. Acta* 240, 463-471.

REUBEN, J. and F. E. POLK (1980): Nucleotide–amino acid interactions and their relation to the genetic code. *J. Mol. Evol.* 15, 103-111.

RIZHKOV, V. L. (1962): Reflexion. In: *The Philosophical Problems of Modern Natural Sciences* (in Hungarian). Akadémiai Kiadó, Budapest.

ROHLFING, D. L. (1976): Thermal polyamino acids: Synthesis at less than 100 °C. *Science* 193, 68-70.

ROHLFING, D. L. (1977): Herrera's sulphobes as protocell models. *2nd ISSOL Meeting,* Kyoto.

ROHLFING, D. L. and S. W. FOX (1969): Catalytic activities of thermal polyanhydro-α-amino acids. *Adv. Catal.* 20, 373-418.

ROOT-BERNSTEIN, R. S. (1982): On the origin of the genetic code. *J. Theor. Biol.* 94, 895-904.

ROSS, H. N. M.; et al. (1981): *J. Gen. Microbiol.* 123, 75.

RÖSSLER, K. (1984): Synthesis of organic molecules by atomic carbon: Relevance to cosmic chemistry. *Lecture in the XXVth COSPAR Congress,* Graz.

RÖSSLER, O. E. (1974a): Chemical automata in homogeneous and reaction-diffusion kinetics. In: M. Conrad et al. (eds.), *Physics and Mathematics of the Nervous System.* Springer-Verlag, Berlin, pp. 399-418.

RÖSSLER, O. E. (1974b): A synthetic approach to exotic kinetics (with examples). In: M. Conrad et al. (eds.), *Physics and Mathematics of the Nervous System.* Springer-Verlag, Berlin, pp. 546-582.

ROSSOW, W. B.; A. HENDERSON-SELLERS and S. K. WEINREICH (1982): Cloud feedback: A stabilizing effect for the early Earth? *Science* 217, 1245-1247.

RUBEY, W. W. (1955): Development of the hydrosphere and atmosphere, with special composition of the early atmosphere. *Geol. Soc. Am. Spec. Pap.* 62, *Crust of the Earth,* 631-650.

SAGAN, C. (1983): The exploration of Saturn's extraordinary moon, Titan. *The Planetary Report* 3, 3-5.

SAGAN, C. and S. F. DERMOTT (1982): The tide in the seas of Titan. *Nature* 300, 731-733.

SAGAN, C. and B. N. KARE (1971): Long-wavelength ultraviolet photoproduction of amino acids on the primitive Earth. *Science* 173, 417.

SAGAN, C. and G. MULLEN (1972): Earth and Mars: Evolution of atmospheres and surface temperatures. *Science* 177, 52-56.

SANCHEZ, R. A. and L. E. ORGEL (1970): Studies in prebiotic synthesis. V. Synthesis and photoanomerization of pyridine nucleosides. *J. Mol. Biol.* 47, 531.

SANCHEZ, R. A.; J. P. FERRIS and L. E. ORGEL (1966a): Conditions for purine synthesis: Did prebiotic synthesis occur at low temperatures? *Science* 153, 72-73.

SANCHEZ, R. A.; J. P. FERRIS and L. E. ORGEL (1966b): Cyanoacetylene in prebiotic synthesis. *Science* 154, 784-785.

SANCHEZ, R. A.;J. P. FERRIS and L. E. ORGEL (1967): Studies in prebiotic synthesis II. Synthesis of purine precursors and amino acids from aqueous hydrogen cyanide. *J. Mol. Biol.* 30, 223-253.

SANCHEZ, R. A.; J. P. FERRIS and L. E. ORGEL (1968): Studies in prebiotic synthesis IV. Conversion of 4-aminoimidazole-5-carbonitrile derivatives to purines. *J. Mol. Biol.* 38, 121-128.

SÁNDOR, I. (1979): Comments to the study of Tibor Gánti entitled: Life Criteria and the Definition of Life (in Hungarian). *Biologia* 27, 97-102.

SAWAI, H. (1976): *J. Am. Chem. Soc.* 98, 7037-7039.

SAWAI, H. (1981): Oligoadenylates formation on an oligouridylate template in the presence of a lead catalyst. *J. Mol. Evol.* 17, 48-51.

SAWAI, H. and L. E. ORGEL (1975): *J. Am. Chem.* Soc. 97, 3532-3533.

SAYGIN, Ö. and E. ELLMAURER (1983): Carbonylphosphate: A phosphorylating agent for adeninenucleotides and carboxylgroups. *4th ISSOL Meeting,* Mainz.

SCHIDLOWSKI, M. (1984): Isotope clues to early organic evolution: Potential and limitations. *XXVth COSPAR Congress,* Graz.

SCHNEIDER-BERNLOHR, H.; et al. (1968): Partial resolution of **DL**-adenosine by template synthesis. *Science* 162, 809-811.

SCHÖPF, J. W. (1976): Are the oldest "fossils", fossils? *Orig. Life* 7, 19-36.

SCHÖPF, J. W. (1983): The early paleontological record. *ISSOL Conference,* Mainz.

SCHÖPF, W. J. and E. S. BARGHOORN (1967): Alga-like fossils from the early Precambrian of South Africa. *Science* 156, 508-511.

SCHRÖDINGER, E. (1944): *What Is Life?* MacMillan Company, USA. Hungarian edition: Selected Studies (in Hungarian) (1970). Gondolat, Budapest.

SCHWARTZ, A. W. (1972): Prebiotic phosphorylation – nucleotide synthesis with apatite. *Biochim. Biophys. Acta* 281, 477-480.

SCHWARTZ, A. W. (1983): Recent developments in the prebiotic chemistry of HCN. *Lecture, 7th ICOL Meeting,* Mainz.

SCHWARTZ, A. W. and G. J. F. CHITTENDEN (1978): Pathways for the prebiotic synthesis of pyrimidines. In: H. Noda (ed.), *Proceedings of the 2nd ISSOL Meeting.* Center for Academic Publications, Japan, Tokyo.

SCHWARTZ, A. W. and S. W. FOX (1967): Condensation of cytidylic acid in the presence of polyphosphoric acid. *Biochim. Biophys. Acta* 134, 9-14.

SCHWARTZ, A. W. and C. PONNAMPERUMA (1968): Phosphorylation of adenosine with linear polyphosphate salts in aqueous solution. *Nature* 218, 443.

SCHWARTZ, A. W.; H. JOOSTEN and A. B. VOET (1982): Prebiotic adenine synthesis via HCN oligomerization in ice. *BioSystems* 15, 191 -193.

SCHWEMMLER, W. (1979): *Mechanismen der Zellevolution: Grundriss einer modernen Zelltheorie.* De Gruyter, Berlin.

SHAPIRO, R. (1983): The improbability of prebiotic nucleic acid synthesis. *4th ISSOL Meeting,* Mainz.

SHEPHERD, J. C. W. (1981): Periodic correlations in DNA sequences and evidence suggesting their evolutionary origin in a comma-less genetic code. *J. Mol. Evol.* 17, 94-102.

SHERWOOD, E. and J. ORÓ (1976): A unified condensation system for the prebiotic synthesis of glycerides, peptides and oligodeoxynucleotides. *Lecture in the 2nd ISSOL Meeting,* Kyoto.

SHIMIZU, M. (1982): Molecular basis for the genetic code. *J. Mol. Evol.* 18, 297-303.

SHORTHILL, R. W.; et al. (1976a): The "soil" of Mars (Viking 1). *Science* 194, 91-97.

SHORTHILL, R. W.; et al. (1976b): The environs of Viking 2 Lander. *Science* 174, 1309-1318.

SIMIONESCU, C.; F. DÉNES and M. MACUVEANU (1973): Synthesis of some amino acids, sugars, and peptides in cold plasma. Electron-microscopic studies on same proteid forms III. *Biopolymers* 12, 237-241.

SIMIONESCU, C.; F. DÉNES and M. TOTOLIN (1981): The synthesis of some lipid-like structures in simulated primeval Earth conditions. *BioSystems* 13, 149- 156.

SIMIONESCU, C.; F. DÉNES and M. TOTOLIN (1983): Nonrandom formation of hydrophobic structures under primeval Earth conditions. *4th ISSOL Meeting,* Mainz.

SINGH, Y. P. (1975): Studies in Abiogenesis of Phospholipids. Ph. D. thesis, Chemistry Department, Allahabad University, India.

SKODA, J. and J. MORAVEK (1966): Formation of uridylyl (3'-5') uridine... by thermic phosphorylation of the corresponding nucleosides with inorganic phosphates. *Tetrahedron Lett.* 35, 4167-4172.

SLABAUGH, M. R.; A. l. HARVEY and J. NAGYVÁRI (1974): The possible role of inorganic thiophosphate as a prebiotic phosphorylating agent. *J. Mol. Evol.* 3., 317-321.

SLEEPER, H. L. and L. E. ORGEL (1979): The catalysis of nucleotide polymerization by compounds of divalent lead. *J. Mol. Evol.* 12, 357-364.

SLEEPER, H. L.; R. LOHRMANN and L. E. ORGEL (1978): Formation of the imidazolides of dinucleotides under potentially prebiotic conditions. *J. Mol. Evol.* 11, 87-93.

SLEEPER, H. L.; R. LOHRMANN and L. E. ORGEL (1979): Template-directed synthesis of oligoadenylates catalyzed by Pb^{2+} ions. *J. Mol. Evol.* 13, 203-214.

SLEMMER, L. (1981): The criteria of life and of the capability of life on the basis of chemoton theory (in Hungarian). *Tájékoztató* (A Műv. Minisztérium kiadványa, Budapest) 1981/2, 44-56.

SLEMMER, L. (1982): A possible way of the formation of prebiotic chemotons by the evolution of self-reproductive microspheres (in Hungarian). *Lecture, XVIth Meeting of the Hungarian Biological Society,* Budapest.

SMITH, A.; C. FOLSOME and K. BAHADUR (1981): Carbon dioxide reduction and nitrogenase activity in organo-molybdenum microstructures. *Experientia* 37, 357-359.

SNYDER, W. D. and S. W. FOX (1975): A model for the origin of stable protocells in a primitive alkaline ocean. *BioSystems* 7, 222-229.

SOFFEN, G. A. (1976): Scientific results of the Viking missions. *Science* 194, 1274-1276.

STEINMANN, G.; A. E. SMITH and J. J. SILVER (1968): Synthesis of a sulfur containing amino acid under simulated prebiotic conditions. *Science* 159, 1108-1110.

STILWELL, W. (1980): Facilitated diffusion as a method for selective accumulation of materials from the primordial oceans by a lipid-vesicle protocell. *Orig. Life* 10, 277-292.

SULSTON, J. E.; et al. (1968): Nonenzymatic synthesis of oligoadenylates on a polyuridylic acid template. *Proc. Natl. Acad. Sci. USA* 59, 726-733.

SZÁDECZKY-KARDOSS, E. (1974): *Geonomy* (in Hungarian). Geochemical Research Laboratory, Hungarian Academy of Sciences, Budapest.

SZATHMÁRY, E. (1982): Unpublished results.

SZATHMÁRY, E. (1984): The roots of the individual organization. In: G. Vida (ed.), *Evolution,* Vol. IV (in Hungarian). Natura, Budapest, pp. 37-157.

SZENTIRMAI, A. (1986): Biotechnology in the chemical industry and biological patents (in Hungarian). *Magy. Kem. Lapja* 41, 361-370.

SZUTKA, A. (1964): Porphine-like substances: Probable synthesis during chemical evolution. *Nature* 202, 1231-1232.

TAULMIN, P., III.; et al. (1977): Geochemical and mineralogical interpretation of the Viking inorganic chemical results. *J. Geophys. Res.* 82, 4625-4634.

TAYLOR, F. J. R. (1974): Implications and extensions of the serial endosymbiosis theory of the origin of eukaryotes. *Taxon* 23, 229-258.

TAYLOR, G. R. (1968): *The Biological Time-bomb.* Thames & Hudson, London.

TAYLOR, G. R. (1970): *The Biological Time-bomb* (in Hungarian). Medicina, Budapest.

TEMUSSI, P. A.; et al. (1976): Structural characterization of thermal prebiotic polypeptides. *J. Mol. Evol.* 7, *105*-110.

TORNABENE, T. G. and T. A. LANGWORTHY (1979): *Science* 203, 51.

TORNABENE, T. G.; et al. (1978): *J. Mol. Evol.* 11, 259.

TRENT, D. J.; R. A. CHASTAIN and A. A. YAYANOS (1984): Possible artefactual basis for apparent bacterial growth at 250 °C. *Nature* 307, 737-740.

TURNER, B. E. (1980): Interstellar molecules. *J. Mol. Evol.* 15, 79-101.

UREY, H. C. (1952a): On the early chemical history of the Earth and the origin of life. *Proc. Natl. Acad. Sci.USA* 38, 351-362.

UREY, H. C. (1952b): *The Planets.* University of Chicago Press, Chicago.

UREY, H. C. (1962): Evidence regarding the origin of the Earth. *Geochim. Cosmochim. Acta* 26, 1-14.

USHER, D. A. (1977): Early chemical evolution of nucleic acids: A theoretical model. *Science* 196, 311-313.

VARELA, F. G.; H.R. MATURANA and R. URIBE (1974): Autopoesis: The organization of living systems, its characterization and a model. *BioSystems* 5, 187.

VOLKENSSHTEIN, M. V. and S. N. FISHMAN (1966): Polynucleotide synthesis at oligomeric templates. *Biopolymers* 4, 77-84.

WAEHNELDT, T. V. and S. W. FOX (1968): The binding of basic proteinoids with organismic or thermally synthesized polynucleotides. *Biochim. Biophys. Acta* 160, 239.

WALKER, G. W. R. (1974): Genetics and the origin of the genetic code. *Orig. Life* 351-356.

WALKER, J. C. G. (1977): *Evolution of the Atmosphere*. MacMillan, New York.

WALKER, J. C. G. (1978): Oxygen and hydrogen in the primitive atmosphere. *Pure Appl. Geophys.* 116, 222-231.

WALSBY, A. E. (1980): A square bacterium. *Nature* 283, 69-71.

WALTER, M. R.; R. BUICK and I. S. R. DUNLOP (1980): Stromatolites 3.400–3.500 Myr old from the North Pole area, Western Australia. *Nature* 284, 443-445.

WARDEN, J. T.; et al. (1974): A re-examination of the zeolite-promoted, clay-mediated peptide synthesis. *J. Mol. Evol.* 4, 189-194.

WATSON, J. D. and F. H. C. CRICK (1953a): A structure for deoxyribose nucleic acid. *Nature* 171, 737-738.

WATSON, J. D. and F. H. C. CRICK (1953b): Genetical implications of the structure of deoxyribonucleic acid. *Nature* 171, 964-965.

WEBER, A. L. (1981): Formation of pyrophosphate, tripolyphosphate, and phosphorylimidazole with the thioester, N,S-diacetylcysteamine, as the condensing agent. *J. Mol. Evol.* 18, 24-29.

WEBER, A. L. (1982): Formation of pyrophosphate on hydroxyapatite with thioesters as condensing agents. *BioSystems* 15, 183-189.

WEBER, A. L. and J. C. LACEY, Jr. (1978): Genetic code correlations: Amino acids and their anticodon nucleotides. *J. Mol. Evol.* 11, 199-210.

WEIMANN, B. J.; et al. (1968): Template-directed synthesis with adenosine-5'-phosphorimidazolide. *Science* 161, 387-389.

WEISS, A. (1981): Replication and evolution in inorganic systems. *Angew. Chem. Int. Ed. Engl* 20, 850-860.

WHITE, D. H. (1980): A theory for the origin of a self-replicating chemical system. Natural selection of the autogen from short, random oligomers. *J. Mol Evol.* 16, 121-147.

WHITE, D. H. (1983): Catalysis of template-directed oligonucleolide formation by small peptides and peptide analogs. *IVth ISSOL Meeting,* Mainz.

WHITE, D. H. (1984): Hydrolytic stability of biomolecules at high temperatures and its implication for life at 250 °C.

WHITE, D. H. and I. C. ERICKSON (1981): Enhancement of peptide bond formation by polyribonucleotides on clay surfaces in fluctuating environments. *J. Mol. Evol.* 17, 19-26.

WHITE, D. H. and M. S. RAAB (1982): A theory for the origin of a self-replicating chemical system ll. Computer simulation of the autogen. *J. Mol Evol* 18, 207-216.

WHITE, D. H.; J. MACKLIN and R. M. KENNEDY (1983): Acyl silicates as activated intermediates in peptide bond formation on clays. *4th ISSOL Meeting,* Mainz.

WHITE, H. B. (1976): Coenzymes as fossils of an earlier metabolic state. *J. Mol. Evol.* .7, 101-104.

WHITE, H. B. (1982): Evolution of coenzymes and the origin of pyridine nucleotides. In: *The Pyridine Nucleotide Coenzymes.* Academic Press, New York.

WILLIAMS, I. R. and J. S. MYERS (1983): Ion microprobe identification of 4.000–4.200 Myr-old terrestrial zircons. *Nature* 304, 616-618.

WOESE, C. R. (1968): The fundamental nature of the genetic code: Prebiotic interactions between polynucleotides and polyamino acids or their derivations. *Proc. Natl. Acad. Sci. USA* 59, 110-117.

WOESE, C. R. (1979): A proposal concerning the origin of life on the planet Earth. *J. Mol. Evol.* 13, 95-101.

WOESE, C. R.; et al. (1966): On the fundamental nature and evolution of the genetic code. *Cold Spring Harbor Symp. Quant. Biol.* 31, 723-736.

WONG, J. T. (1975): The evolution of universal genetic code. *Proc. Natl. Acad. Sci. USA* 72, 1909-1912.

WONG, J. T. (1976): A co-evolution theory of the genetic code. *Proc. Natl. Acad. Sci. USA* 73, 2336-2340.

ZANG, A. J.; P. l. GRABOWSKY and T. R. CECH (1983): Autocatalytic cyclization of an excised intervening sequence RNA is a cleavage-ligation reaction. *Nature* 301, 578-583.

ZUCKERMANN, B. (1977): Interstellar molecules. *Nature* 268, 491-495.

YANAGAWA, H. and F. EGAMI (1980): Formation of organized particles, marigranules and marisomes from amino acids in a modified sea medium. *BioSystems* 12, 147-154.

YATKO-MANZO, E.; F. RODDY; R. G. YOUNT and D. E. METZLER (1959): The catalytic decarboxylation of pyruvate by thiamine. *J. Biol. Chem.* 234, 733-737.

YBANEZ, J. D.; A. P. KIMBALL and J. ORÓ (1971a): Condensation of mononucleotides by imidazole. *J. Mol. Evol.* 1, 112-114.

YBANEZ, J. D.; A. P. KIMBALL and J. ORÓ (1971b): Possible condensation of mononucleotides by cyanamide. *Science* 173, 444.

YOUNT, R. G. and D. E. METZLER (1959): Decarboxylation of pyruvate by thiamine analogues. J. *Biol. Chem.* 234, 738-741.

YUEN, G. U.; L. G. LAWLESS and E. H. EDELSON (1981): Quantification of monocarboxylic acids from a spark discharge synthesis. *J. Mol. Evol.* 17, 43-47.

Index

A

acetylphosphate 352
active site 412
adaptation process 463
aldocyanine microspheres 330
Alfvén 306
alive 278
alkyl ethers 345
Altman 399
amino acids 315
aperiodic crystal 264
aperiodic solid body
  concept of 264
apoenzyme 409
Astbury 465
autogenetic model 480

B

Bahadur 293, 313, 329, 332, 502, 504, 507
Baross 387, 503
basic acting principle 274
Bauer 259, 261
Békés 291
Bell 465
Benard phenomenon 385
Bernal 396
Bernard 256
Bertalanffy 259, 266
Berthelot 311
Berzelius 255
biogenesis
  conditions of 384
bioid 293
biological evolution 311, 391
biotechnology 493
  "classical" 494
  "new" 494
black smoker 387, 451
breathing mechanism 415
Brewin 399, 419
Brittain 388
Bungenberg de Jong 328
Buvet 278, 397

C

Callisto 449
canalizing effect 425
carbohydrate metabolism 341
carbohydrates 316
catalysis
  primitive 396
catalytic activities 397
catalytic superregulation 397
Cech 399
central dogma 419
central nervous system 280
Chargaff's rule 368, 464
chemical evolution 311
chemical networks 270
chemical work 269
chemoton 334
  prebiotic 338, 356
  primitive 341
chemoton model 251
chemoton theory 454
coacervates 328, 331, 505
code
  origin of 414, 422

coded information 456
codon triplets 423
coenzyme 397, 409
coenzyme regeneration cycles 499
coenzyme-type compounds 397
complementary pairing 414, 468
concept of life 251
connective subsystems 346
conservative sections 423
constant nonequilibrium 275
Corliss 387
Cramer 344
Crick 265, 399, 464
Csányi 480
Csányi's theory 479, 482
cyan polymers 324
cybernetic system 274
cycle stoichiometry 251

D

dead 278
Decker 293, 316, 393
dehydration 278
Deming 503
Dénes 330
deoxyribonucleic acid 257
diad 489
dicarboxylic acids 316
dirty ice 305
divine machine 260
DNA molecule 265
DNA-based metabolism 472
Draganic 313
dry celestial bodies 437
dust particles 307

E

Eigen 293, 335
electrical discharge 312
endosymbiosis 488, 490
Engels 255
entropy 266
enzymatic regulation 396
enzyme activity
  appearance of 396
enzyme RNA 400, 402
e-RNA 405
eukaryote 287
  evolution of 491
Europa 449
evolution 286, 479
evolutional history 458
excitability 284
exobiology 427
extraterrestrial life 427

F

fatty acids 320
Fischer 255
fluid automata 273, 283
Folsome 329, 332, 388, 503, 507
Folsome-Brittain experiment 388
formaldehyde 341
formose reaction 316, 341
Fox 293, 324, 329, 332, 387
Fraser 329, 332, 503
frozen accident 467
Furberg 465

G

Galilean moons 438, 449
Ganymede 449
Garrison 311
gas exchange experiments 445
GC-MS method 443
genealogical trees 488
general evolution theory 482
genes
  origin of 414

genesis
  of Earth 300
  of life 298, 331, 385
genetic code 332
genetic material
  structure of 464
giant planets 452
globular RNA 405
glyceraldehyde 342
glycerol diphythanol ether 345
Greenberg 305
growth 285
Gyarmati 263, 272
Gyarmati-Li theory 272

H

Haldane 507
Halmann's reaction 346
Hartman 286
Harvey 399
hereditary change 286
  adaptive 464
hereditary changes 457
Herrera 433
Hevesi 254
Hidvégi 345, 351, 410
Hofmeister 255
homeostasis 284, 290
homogeneous pairing 414
homunculus 501
hybrid cells 494
hydrocarbons 320
hypercycle 334
  model 293
  theory 335

I

imidazole 322, 344
individuality 287
information carrier 344
information subsystem 344
informational operation 285
information-carrying subsystem 288
inherent entity 283
inherent stability 284, 288
inherent units 289
internal control 288
interstellar molecules 301
inverse assimilation 393
  of cycle 342
Io 438
isoprenoids 321
Ivanovskii 252

J

Jacob 258, 274
Jeewanu 329, 332, 504
Jupiter 438, 452

K

Kenyon 330
Keosian 502
Khorana 344
Kimball 317
kinetic regulation 273
Korányi 399, 410
Kuhn 293, 333, 403

L

label release experiments 446
Langenbeck 316
Laplace's theory 306
Leduc 434
Leibniz 259
Lengyel 272
Liebig 255
life
  carbon based 434
  chemosynthetic living world 451
  non-carbon based 429

particles of 504
silicon-based 429
synthesis of 500
life criteria 278, 282
absolute (real) 282, 283
potential 282, 285
life phenomena 282
limit cycle 290
lipid-like compounds 320
living
capable of 278
living systems
synthesis of 500
lo 449
Loeb 311
loop-RNA 405

M

Margulis 489
marigranules 330
marisomes 330
Mars 440
Martian day 440
Matthews 324
melanoidin microspheres 330
membrane
building units of 345
synthesis of 359
membrane potential 329
membranogen molecules 359
Mercury 437
metabolic network 270, 338
primitive 341
metabolism 254, 283, 287, 290
microfossil 299
microscopic structures 328
microspheres 324, 329
Miller 256, 298, 311
minimum system 283, 293
Moiseeva 340
moment of death 280
monad 489
monas 260
monocarboxylic acids 316
Morowitz 438
mortality 286
Moser 324

N

natural automaton 260
negative
copy 414
strand 414
neobiogenesis 502
Neptune 452
Neuberg 316
Nissenbaum 330
non-enzymatic replication 404
non-enzymatic template polymerization 325
non-living 278
nucleoside imidazoles 344
nucleotide bases 317
nucleotide imidazolides 325, 327, 410

O

oldest rocks 299
Onsager 263
Oparin 294, 328, 331, 502
organisational hierarchy 493
origin of life 297
Oró 317
overall equation
of prebiotic chemoton 378

P

pairing vocabulary 423
palindromic scqucncc 414
paradigm theory 495

Pasteur 502
phosphorus-containing compounds 318
phosphorylating agent 318
photosynthesis 436
planetesimals 307
polydiad 489, 492
polymerase function 419
polytriad 489, 492
prebiotic
  condition 297
  evolution 311, **383**, 391
  reaction 298
Prigogine 263
primitive atmosphere 309
primitive Earth 300
primordial Earth 298
principle of constant non-equilibrium 262
principle of life
  by Bauer 262
program-control 285
prokaryote 287
prokaryotic
  cell 470
  minimum model 474
  minimum system 478
proliferating microspheres 455
proliferation 285
property-determining factor 458
prosthetic groups 409
proteinoid microspheres 330, 505

proteinoids 324
  thermal 324
proteins 255
pyrolytic release experiments 444

Q

quantitative biology 453
quantitative theoretical biology 383
quasi-species 337

R

Ranganayaki 502, 504
Raulin 435
replication
  identical 483
  non-identical 483
reproductive systems 481
reverse transcriptase 420
Reynolds 451
ribonucleotide imidazolide 344
ribose 342
RNA
  evolution of 400
RNA-polymerase
  origin of 417

S

Sagan 438
Saturn 452
Schidlowski 300
Schopf 299
Schrödinger 259, 264, 454
Schröndinger principle 454
Schwartz 325
Schwemmler 489, 492
selective permeability 329
sensitivity 284
sequence-reading mechanism 457
Serial Endosymbiosis Theory 489
Simionescu 330
Slemmer 279, 390
small-molecular products 315
soul
  according to Leibniz 259
spirit 274
square bacterium 253
Squyres 451

Stanley 252
stellesimals 306
stoichiokinetic coefficient 375
stoichiokinetics 374
stoichiometric
  organisation 493
  regulation 271
Strecker synthesis 315
stromatolite 299
sugar phosphates 317
sulphur seas
  of Io 439
supracellular organization 288
symbiontes 491
symbiosis 487
synthese
  by reaction networks 498
  one-step 496
system precursor 481
Szádeczky-Kardoss 385
Szathmáry 335, 356

T

Taylor 489, 492, 500, 501
template effect 327
template polymerization system 292
template synthesis 361
terrestrial metabolism 436
theoretical biology
  by Bauer 261
  exact 251
Titan 439
Todd 344
total synthesis 502
Trent 503
triad 489
turnover number 499

U

ultraviolet radiation 312
unit of life 267
Uranus 452
Urey 309, 311

V

Venus 437
Viking-1 440
Viking-2 440
virales 252
virus 252
vis vitalis 254, 274
von Neumann 455
Voyager spacecrafts 449

W

Walsby 253
Watson 265, 464
Watson-Crick model 454
White 410
Woese 299, 385
Woodward's reagent 344

Y

yellow stuff 306

A

α-state 279

B

β-state 279, 291

Γ

γ-state 279